Getting Trains into Service

Conference Organizing Committee

I Arthurton
Consultant

D Barney
The Engineering Link

L Bartholomew
Serco Railtest Limited

D Gillan
Railway Industry Association

R Goodall
Loughborough University

R Goulette
Adtranz

K Herriman
Adtranz

R Jones
Virgin Trains

S Maltby
Railway Division Young Member

C May
Rail Technology Services

C Perry (Chairman)
AEA Technology Rail

J Smith
Anglia Railways

T Taig
AEA Technology plc

J Wragg
Interfleet Technology Limited

International Liaison

Tony Roche FEng

Honorary Committee

R Aylward
Forward Trust Rail Limited

A Baker
Angel Train Contracts

I Braybrook
English, Welsh, and Scottish Railway Limited

C Garnett
Great North Eastern Railway Limited

M Lloyd
ALSTOM Transport Limited

P Staehr
ABB Daimler-Benz Transportation

I Warburton
Virgin Rail Group Limited

P Watson OBE
AEA Technology plc

D Wilson
Siemens Transportation Systems

IMechE
Seminar Publication

Getting Trains into Service
International Railtech Congress '98

24–26 November 1998
National Exhibition Centre,
Birmingham, UK

Organized by the Railway Division of the
Institution of Mechanical Engineers (IMechE)

In association with the:

Institution of Railway Signal Engineers
Institution of Civil Engineers
Railway Industry Association
Unione Internationale des Transport Publics
Institution of Electrical Engineers
UNIFE
Japan Society of Mechanical Engineers

IMechE Seminar Publication 1998–21

Published by Professional Engineering Publishing Limited for
The Institution of Mechanical Engineers, Bury St Edmunds and London, UK.

First Published 1998

This publication is copyright under the Berne Convention and the International Copyright Convention. All rights reserved. Apart from any fair dealing for the purpose of private study, research, criticism or review, as permitted under the Copyright, Designs and Patents Act, 1988, no part may be reproduced, stored in a retrieval system, or transmitted in any form or by any means, electronic, electrical, chemical, mechanical, photocopying, recording or otherwise, without the prior permission of the copyright owners. Reprographic reproduction is permitted only in accordance with the terms of licences issued by the Copyright Licensing Agency, 90 Tottenham Court Road, London W1P 9HE. *Unlicensed multiple copying of the contents of this publication is illegal.* Inquiries should be addressed to: The Publishing Editor, Professional Engineering Publishing Limited, Northgate Avenue, Bury St. Edmunds, Suffolk, IP32 6BW, UK. Fax: 01284 704006.

© The Institution of Mechanical Engineers 1998

ISSN 1357–9193
ISBN 1 86058 186 2

A CIP catalogue record for this book is available from the British Library.

Printed and bound in Great Britain by Bookcraft (Bath) Limited.

The Publishers are not responsible for any statement made in this publication. Data, discussion, and conclusions developed by authors are for information only and are not intended for use without independent substantiating investigation on the part of potential users. Opinions expressed are those of the Author and are not necessarily those of the Institution of Mechanical Engineers or its Publishers.

Related Titles of Interest

Title	Editor/Author	ISBN
New Trains for New Railways	IMechE Conference 1998–7	1 86058 146 3
Train Maintenance Tomorrow and Beyond	IMechE Conference 1997–1	1 86058 095 5
Better Journey Time – Better Business	IMechE Conference 1996–8	0 85298 997 0
Implementing Rail Projects	IMechE Conference 1995–1	0 85298 948 2
Fault Free Trains – A Reality	IMechE Seminar 1998–8	1 86058 133 1
Railway Traction and Braking (Railtech '96)	IMechE Seminar 1996–19	1 86058 018 1
Design, Reliability, and Maintenance for Railways (Railtech '96)	IMechE Seminar 1996–18	1 86058 017 3
*Railway Rolling Stock (*Railtech '96*)*	IMechE Seminar 1996–17	1 86058 016 5
Railway Engineering, System and Safety (Railtech '96)	IMechE Seminar 1996–16	1 86058 015 7

For the full range of titles published by Professional Engineering Publishing contact:

Sales Department
Professional Engineering Publishing Limited
Northgate Avenue
Bury St Edmunds
Suffolk
IP32 6BW
UK

Tel: 01284 724384
Fax: 01284 718692

Professional Engineering Publishing

Professional Engineering Publishing Limited is the new name for the publishing house of the Institution of Mechanical Engineers.

The reasons that we have changed our name after 25 years, from Mechanical Engineering Publications, are quite simple. As our readership and publishing aims grow, we need a name which reflects our plans for the future and our wider audience. Engineering is becoming a more complex and involved discipline; the boundaries between the various branches are becoming progressively blurred. There are increasing publishing opportunities for all our areas of interest including Journals, Newsletters, Electronic Products, Books and Magazines.

Professional Engineering Publishing is a leading international engineering publisher, the publishing house of the Institution of Mechanical Engineers (IMechE) and the exclusive European Agent for The American Society of Mechanical Engineers (ASME).

Professional Engineering Publishing will serve the international engineering community by expanding our publishing activities for the professional reader.

The history of mechanical engineering
A past that has built our future

An Engineering Archive

A Selection of Papers From the Proceedings of the Institution of Mechanical Engineers

Edited by Professor Desmond Winterbone
Archivist Keith Moore

An Engineering Archive celebrates the 150th anniversary of the foundation of the Institution of Mechanical Engineers and offers an excellent opportunity to look back at the contribution made by the Proceedings of the Institution to the whole area of mechanical engineering.

This volume contains a selection of papers, dating back to 1848, which have been chosen from the first 100 years of the Proceedings to exemplify both landmark accounts of engineering technology and science, and contributions from famous authors.

The papers reproduced are in facsimile, to show their original form. A number of the most memorable plates are included together with biographies of key engineering innovators.

300x210mm / 648 pages / 1997
Leather Bound:
ISBN 1 86058 052 1   £95.00
Quality Hardcover:
ISBN 1 86058 053 X   £59.00

Progress through Mechanical Engineering

By John Pullin

Published to mark the 150th anniversary of the founding of the world's first mechanical engineering institution, this book, with a foreword by HRH The Prince of Wales, charts the progress of the profession and the influences that have shaped engineering. At the same time it shows the key role that engineers and engineering have had, and will continue to have, on the world we live in, and the quality of life we lead, and illustrates the part that the Institution of Mechanical Engineers has played in those developments.

This volume will appeal to professional engineers, academics, those involved with the history and philosophy of science as well as readers with a more general interest.

1899163 28 X / 292x197mm / Hardcover / 288 pages / 1997
£28.00

Note: free delivery in the UK. Overseas customers please add 10% for delivery.

Credit card orders welcome

For further information:
Telephone Hotline - (24 Hour Service) - +44 (0) 1284 724384
Orders and enquiries to:
Sales Department, (eng), Professional Engineering Publishing, Northgate Avenue, Bury St Edmunds, Suffolk, IP32 6BW, UK.
Fax: +44 (0) 1284 718692   E-mail: sales@imeche.org.uk

Professional Engineering Publishing

New Book Titles

Designing Cost-Effective Composites

This volume, based on the international conference held in London, UK, 15-16 September 1998, presents papers that are drawn from a range of experience and research into the uses, development, and analysis of composites.

It is an invaluable volume for design engineers, materials scientists, technologists, engineers in the fields of manufacturing, power and process industries, aerospace, automotive, and marine engineering. Researchers in academe, as well as those in government or independent laboratories, will also find this a useful source of information.

1 86058 148 X / 234x156mm / Hardcover / 291 pages / 1998 / £69.00

Credit card orders welcome

Note: free delivery in the UK. Overseas customers please add 10% for delivery.

Forging and Related Technology (ICFT '98)

This IMechE conference volume addresses the fundamental and practical issues which are crucial to the successful production and optimum utilization of forged parts on a global basis. It is essential reading for all forward-looking manufacturers, designers, value analysers, material technologists, and users in the forging supply and related industries.

1 86058 144 7 / 234x156mm / Hardcover / 464 pages / 1998 / £124.00

Design Reuse
Engineering Design Conference '98

Edited by Dr S Sivaloganathan and Dr T M M Shahin

This volume containing appers presented at a conference organised by Brunel University in June 1998.

Drawing on the diverse experience of different sectors of industry and academic research, this prestigious collection of papers examines the ways in which the expertise in design can be structured and reused to carry the design process forward.

1 86058 132 3 / 234x156mm / Hardcover / 732 pages / 1998 / £189.00

Professional Engineering Publishing

For further information:
Telephone Hotline - (24 Hour Service) - +44 (0) 1284 724384
Orders and enquiries to:
Sales Department, Professional Engineering Publishing,
Northgate Avenue, Bury St Edmunds, Suffolk, IP32 6BW, UK.
Fax: +44 (0) 1284 718692   E-mail: sales@imeche.org.uk

Contents

C552/007/98	**The Toronto transit commission's T1 Subway car** I A Johns	1
C552/043/98	**The key to successful, cost-effective, independent safety audit** S Beech	35
C552/081/98	**How MM can add value to the business** M J Hudson	41
C552/031/98	**Whither QRA?** D Edge and W Harrison	59
C552/037/98	**Group standards and vehicle acceptance** A Sutton	69
C552/011/98	**Commissioning trains** R Hardi	81
C552/096/98	**Reliability growth in the rail industry** M H Irwing	87
C552/016/98	**The safety consideration of introducing new rolling stock on an adjacent railway system** W Hoskins	101

C552/007/98

The Toronto transit commission's T1 Subway car

I A JOHNS BSc, PEng, CEng, MIMechE
Bombardier Inc., Ontario, Canada

The Bombardier-built T1 subway car, with microprocessor control of several major on-board systems, is the latest addition to the Toronto Transit Commission's fleet.

First pioneered in 1962, the design of such seventy-five foot long subway cars has since then been refined jointly by the Commission, Canadian car manufacturers and suppliers of vehicle equipment systems worldwide.

The paper describes the evolution and design of this recently delivered car, a car in many ways fully representative of modern North American subway car practice.

1 HISTORICAL

The population explosion after the last war produced a demand for public transport in Toronto which greatly overloaded the existing system and, in conjunction with the increased use of private automobiles, produced chronic congestion in down-town areas at peak periods (ref. 1).

The street cars in use in Toronto before the opening of the first subway were PCC cars with frame-mounted motors and right-angled drives. As the Toronto Transit Commission (TTC) were favourably disposed with the quietness and operation with this type of drive, it was specified for the cars of Canada's first subway line, the Yonge Street line which opened to traffic in March 1954. The adopted track gauge was that of the surface lines, 1 495 mm (4 ft 10 7/8 in), as it was hoped that the streetcars might share the downtown subway tracks. As the stations were to have platform lengths of 152.4 m (500 ft), a car 17 374 mm (57 ft) long was designed so as that eight car trains could be accommodated.

Thanks to the market forces then in existence, the first subway cars were manufactured in England by the Gloucester Railway Carriage and Wagon Company. The car was designated class "G", had all axles motored but unlike the PCC cars did not have rheostatic braking. Married-pair operation was used permitting four and six car trains during the off-peak hours. Anticipating future developments, eight aluminium subway cars were subsequently ordered from Gloucester which saved about eleven thousand pounds per car over the earlier-built steel cars.

The thirty-six additional subway cars acquired by the TTC in 1962-3 were in many ways as revolutionary as the PCC cars. After a careful review of subway tunnel dimensions and tolerances, Hawker Siddeley Canada Ltd. showed the TTC that six, 22 860 mm (75 ft) cars could do the work of eight of the existing cars, with all the attendant savings that would accrue by having less equipment.

Fortune does not always favour the brave. In response to the TTC's Request for Proposals for such new cars, a contract was awarded to the Montreal Locomotive Works. These new class "M" cars (ref. 2) pioneered the design and development of the seventy-five foot long subway car, a design now in global use, with such features as:

- Main structure of full-length, one piece aluminium extrusions
- Highest load/tare ratio of any rapid transit car
- Lowest weight per linear or per square foot of any rapid transit car
  For the Commission, the new equipment included:
- Rheostatic service brake
- Load weighing for performance control
- Dual performance mode
- Waste heat recovery
- Fluorescent lighting with 400 Hz AC power from a motor-alternator set

As Hendry notes (ref. 3), the adoption of rheostatic braking on the class M-cars was de rigeur as enormous quantities of brake shoe dust had been spread throughout the original tunnels and stations by the G-cars. The intent behind the dual performance mode was for newer cars to have a low performance when running with the class "G" cars and high performance when running alone.

Subway expansion dictated further orders of subway cars and by 1975 there were an additional three hundred and twenty-eight cars of three different classes with progressive design refinements. The later cars built by Hawker Siddeley Canada Ltd. were designated classes "H1", "H2" and "H4". Noteworthy was that all the class M- and H- cars could be trained in multiple unit in any consist notwithstanding that there were propulsion controls from two different manufacturers and traction motors of three different types. The practise of having all axles motored continued as did married pair operation with the batteries on the A-cars, the air compressors on the B-cars.

One important intermediate development milestone was re-equipping six H2 cars in 1973 with Hitachi chopper controls with electro-dynamic regenerative braking to evaluate their operational, energy and maintainability potential. Such was the satisfactory experience with these cars (designated H3 conversion cars) that in September 1974 (ref.4), the Commission allowed that chopper type electronic controls could be offered as an alternative to camshaft controls for vehicles for the new Spadina and Bloor-Danforth line extensions. The subsequent evaluation of the tenders with respect to energy and capitalized maintenance costs resulted in chopper controls being selected for the H5 cars.

While outwardly similar to their predecessors (with which they were to operate in multiple with equivalent performance), the H5`s boasted quite a number of important new features:

- Chopper propulsion controls
- Separately-excited traction motors
- Regenerative braking with blended friction braking
- Air conditioning
- AC power for auxiliary systems
- Collision posts and anti-telescoping beams
- Cantilevered seating (to improve car cleaning)

Although Hitachi had pioneered the application of choppers on the TTC subway cars, Garrett Manufacturing Ltd. provided the first fleet of such equipment.

Eight class H6 cars were delivered in late 1976. From the outset, however, a number of quality and technical problems beset these vehicles. It was only after extensive type tests, a series of retrofits and the inevitable disruptions at the manufacturing plant that the vehicles were finally delivered.

Four additional vehicles had to be added onto the initial order to replace cars which had been vandalized. Although all the class H- cars had cast steel trucks of a design progressively refined by the Canadian steel manufacturer Dofasco, these four vehicles were equipped with a fabricated truck of German design to which the Carbuilder, Hawker Siddeley Canada Ltd., had acquired the North American manufacturing rights under a licensing agreement. Unlike the existing trucks which had bolsters and sliding wear plates, these new trucks supported the carbody directly, a design common in Europe but virtually untried in North America.

From November 1980, the TTC began to run all lines with cars on low performance. This produced a remarkable improvement in the reliability of motors and associated control equipment (ref. 3).

At the end of December, 1983 the TTC awarded a contract to the Urban Transportation System Corporation (UTDC) for one hundred and twenty-six cars, designated class H6. These cars too had a number of new features including:

- Brush chopper propulsion equipment (developed from the Garrett H5 system)
- Force-ventilated traction motors
- Separate low voltage power supply and battery charger

- Overhead as well as baseboard heating
- Fabricated trucks made under license in Canada

These cars were afflicted with similar problems to those of the class H5. As type and performance tests progressed, deliveries had to be halted and cars stored at the manufacturing plant. Inevitably, the G-cars had to soldier longer on in service than intended with all the disruption that this brought. In time, however, both the class H5 and H6 bettered the reliability of the earlier cars with cam-controlled propulsion equipment.

Whereas the eight prototype fabricated H5 trucks had little impact on the Commission's running shop maintenance practices, a fleet of vehicles with such trucks was a maintenance nightmare. Each time a truck had to be replaced, no less than eight mechanical links had to be disconnected. Although the H6 vehicles rode superbly, to say that they were unpopular with depot staff is an understatement.

Moreover, UTDC were about to embark on a Build-Operate-Transfer programme in Ankara, Turkey with a version of the H6 car, the first export model of such a car, and were well aware that the maintenance shortcomings of the existing truck created much risk. Out of the shortcomings of this welded truck came the T1 Prototype car.

2 T1 PROTOTYPE CAR

Toronto uses much salt on its streets during the long Canadian winter. This gets tracked into the cars by the travelling public and, inevitably, finds its way beneath the door apertures and floor covering.

From the class H2 cars and up, steel channels were nested inside the aluminium side sills below the door apertures to carry the shear loads across the aperture.
Careful design and good sealants notwithstanding, salt from the streets can often reach the aluminium-to-steel faying surfaces; in time, some aluminium side sills corrode to the point where the load-carrying capability of the sill is often impaired. Replacement of part, or all, of the sill is then necessary, a task which the TTC depot staff handle most adeptly.

H5 car #5796 was such a car. Once it had been disassembled, a section of the car was unofficially mocked-up with longitudinal seating. At this point, the UTDC formally approached the Commission to ask that if they were to complete the car interior on the TTC's behalf with longitudinal seating, might the car be equipped with their newly-designed fabricated truck they were anxious to prove for the Ankara project. The resulting truck, essentially a fabricated version of the classic North American cast steel transit truck, had a bolster so as to overcome the inordinate amount of time to remove and install trucks.
And thus the T1 Prototype (T1P) vehicle was born, a test bed for prototyping new ideas from the Commission and equipment vendors alike. For example, long-stroke pneumatic operators replaced rotary electric door operators, vandal-resistant upholstered insert seats replaced FRP seats. The fabricated truck featured components new to the TTC such as compact tread brake units (TBU's) of European origin with spring-applied parking brakes and a new light-weight current collector used on many transit properties in the U.S.A..

Not all new ideas were adopted: no sooner had the T1P car re-entered service to a blaze of publicity, than public pressure persuaded the TTC to revert back to a traditional mixture of longitudinal and transverse seating; following push-through trials, the parking brake portion of the compact TBU was disabled as the smell of "hot brakes" was disconcerting to motormen always alert to the possibility of undercar fires; and, quite simply put, the TTC preferred their existing current collector. (All these features were, out of interest, adopted on the Ankara vehicle, the first export model of a TTC-type car).

Other married pairs subsequently came under the umbrella of the T1P programme. Advanced static convertors made by (then) NEI Control Systems were subsequently installed on class H6 cars #5932/3 and an AEG-Westinghouse Advanced Electric Cam propulsion system on class H4 cars #5604/5.

The car to which the T1P car was married, car #5797, was also equipped with brand new Dofasco cast steel trucks (with bolsters) which later resulted in a series of comparative ride quality tests between the two vehicles.

Unquestionably, the prototype testing helped the TTC to draft the technical specifications for the next generation of subway cars. However, as will be learned, some of the most important systems on the T1 car such as the friction brake and propulsion systems were never assessed under TTC conditions before award of contract.

3 THE TTC T1 SPECIFICATION

In the Contract Documents for the Supply of T1 Subway Cars (Proposal number: P05D-9095), the Commission left the Prime Contractor free to put forward his own detailed design of subway car subject to compliance with the specification.

Parallel specifications were written permitting the Carbuilder to propose either camshaft-control or AC propulsion and either an electro-pneumatic single pipe air brake system with an electric emergency brake loop continuity circuit or the classic electro-pneumatic brake system with an emergency brake pipe.

Overall, the specification stressed that the proposed equipment had to be of a proven design with a guaranteed high degree of reliability and, moreover, complex equipment had to have self-diagnostic capabilities. More personnel training than in recent delivery programmes was specified and the delivery of portable test equipment and maintenance manuals was linked to achievement of milestone payments.

Long before Canada joined the North American Free Trade Association (NAFTA), the earlier TTC specifications had always referred to American jurisdictional standards from such bodies as A.S.T.M., A.I.S.I., A.A.R., A.W.S. and S.A.E. and, not unexpectedly, these were retained in the T1 specification. For the first time on a TTC subway car delivery programme, the T1 vehicle had to demonstrate compliance with the FRA`s "Recommended Fire Safety Practices for Rail Transit Materials Selection"(Federal Register, Volume 54, No.10, January 17th, 1989).

As on previous car delivery programmes, there was a negotiated contract between the TTC and a Canadian carbuilder for the supply of T1 cars. If in the past, cars had been delivered to the TTC by HSCL and UTDC, the sole remaining Canadian carbuilder was now Bombardier of Montreal.

Bombardier, although Prime Contractor, make certain that the system suppliers accept much of the inherent risk (perhaps more risk than they have assumed hitherto on contracts for the TTC).

The base bid submitted by Bombardier was to equip the first seventy-two cars of the potential two hundred and sixteen car order with the Advanced, Electric Cam control system as used on the Metropolitan Transit Authority - New York City Transit. The intent was to use service-proven equipment until such time as the reliability of the AC controls improved.

However, after exhaustive discussions with both Bombardier and ADtranz, the TTC selected AC propulsion from the first car and up. Moreover, with the prospect of less pneumatic components and fewer rotating machines to overhaul in the future, the TTC chose a single pipe air brake system and static invertors. Unlike most other North American transit authorities, the TTC do not hire engineering consultants, there being an enormous amount of technical expertise in-house.

(In retrospect, the choice of AC propulsion is hardly surprising seeing as in recent summers power "Brown Outs" are common in Toronto thanks to the electrical demand of domestic air conditioning units. To be seen adopting traction equipment which was anything less than highly energy efficient would not have been politically correct for either the TTC or the Government of Ontario).

In the past, the TTC had accommodated development work on their property of advanced solid state equipment on the H5 and H6 vehicles. Such was the disruption often caused within the TTC`s maintenance depots by vehicles without Final Acceptance Certificates (i.e., transfer of ownership), the granting of Preliminary Acceptance Certificates allowing T1 vehicles to be shipped from the Bombardier Thunder Bay, Ontario manufacturing facility was made far more stringent. This dictated that the facility first become certified to ISO 9002. (And now marching in step with the rest of the Bombardier organization, the facility is presently being audited with the intent of becoming certified to ISO 9001).

The newly-revamped test track at Thunder Bay permits far more performance testing and charting to be made, including regenerative braking and rail gap testing, than was hitherto carried out on the H1 to H6 cars.

As specified, two prototype cars with unfinished carbody shells and only essential equipment were sent to Toronto for advanced testing of E.M.I. .
Six, pre-production cars followed some eight months after the delivery of the prototype cars for exhaustive performance and reliability testing. These six cars were accepted by the TTC on March 8, 1996 and first operated in revenue service on March 11, 1996.

Figure 1. General View of TTC T1 Car.

The General Arrangement and Principal Dimensions of the T1 cars are given in Appendix 1.

4 RELIABILITY

The TTC's specified reliability requirements for the major sub-systems of the T1 car and for the car as a whole are tabulated in Table 1 below. The specified levels are based on the standard acceleration rate with occasional operation in high rate.

Bombardier had to guarantee that these requirements would be met during the Reliability Verification Programme, which begins once each vehicle has been in service for five hundred miles. For contractual reasons, three groups each consisting of seventy-two cars are defined.
The T1 specification states that the reliability performance shall be measured using the unit "Mean Distance Between Relevant Failures" (MDBRF), where:

$$\text{MDBRF} = \frac{\text{Total travelled subway car miles}}{\text{Total number of relevant failures}}$$

A relevant failure being a failure which requires a <u>non-scheduled</u> maintenance action such as repair or replacement and that is not:

- a consumable item
- a dependent failure
- caused by a collision, an accident or negligence of the Commission
- caused by improper maintenance practises of the Commission
- a preventive maintenance action

Such a non-scheduled maintenance action can either be made in revenue service (by travelling Line Mechanics, for example) or at the depot. The numerator used to calculate the MDBRF is the total revenue fleet mileage of all the subway cars in that particular subway car group.

The allocation of MDBRF is based on the simple formula:

$1/f = 1/f1 + 1/f2 + 1/f3$ etc.

where f is the overall MDBRF for the car or group and f1, f2 etc. are the individual MDBRF's for the group or component. The allocation of MDBRF's within a given group is the responsibility of the Carbuilder. (With new equipment, which is not necessarily service-proven on other transit properties, the Suppliers may often sign up for requirements which they do not know if they can readily meet).

This Reliability Verification Programme is supported by a rigorous Failure Reporting and Corrective Action (FRACAS). The key elements of this procedure for collecting data about failures, analysing the failures and forcing resolution and implementation of corrective action is:

- The Commission perform the car repairs and document them in the TTC's Subway Maintenance System (SMS) database
- Bombardier and the Commission hold a weekly relevancy meeting to review all the repairs performed on cars and to decide which should be considered failures chargeable against the reliability of the car
- All relevant failures are reviewed by Bombardier and are classified into Investigation Items, regrouping failures by problem areas or trends. For each item, a responsibility is assigned to investigate the problem, find the root cause and identify the corrective action to be taken.
- Whenever required, a detailed Failure Analysis Report is requested from a sub-system supplier to thoroughly document the cause of a given failure of a particular piece of equipment and to provide valuable input about that Investigation Item.
- The status of all investigations as well as the performance of each sub-system are reviewed weekly to monitor progress and put pressure on the resolution progress.

The vehicle sub-systems defined in Table 1 and the overall subway car shall be deemed as having successfully passed the Reliability Verification Programme when the relevant twenty-six week moving average MDBRF has met its specified level for all but a maximum of seven weeks during a twenty-six consecutive week period. In addition, the average of the twenty-six consecutive week period must pass the specified level.

At the time of preparing this paper for publication (98 Aug 22), one hundred and thirty T1 cars (all seventy-two cars of Group 1 and forty-eight cars of Group 2) have been delivered and are in-service with the TTC as scheduled. The graph presented in Appendix 2 of this paper shows the Overall Car twenty-six moving window MDBRF for car Group 1 as of this date.

This graph hints at the effort and time needed to get a reliability growth sufficient to meet the specified MDBRF requirements. The linearity of this curve stems from the fact that each MDBRF measurement represents twenty-six weeks of data. While this eliminates any short term volatility, the drawback is that any problem or trend is retained for a long time thus reducing the slope of the reliability growth curve. This growth can only be sustained by the constant resolution of problems and the implementation of appropriate corrective actions (retrofits).

TABLE 1 - RELIABILITY OBJECTIVES

SUB-SYSTEM	DESCRIPTION	MINIMUM MDBRF (MILES)
Subway Carbody	Body work, roller curtain destination sign, seats, flooring, run number sign, windows, battery tray, end and cab doors.	200,000
Truck & Suspension	Complete truck except traction motors, brake units, gear unit and current collection equipment.	300,000
Main Auxiliary	Inverter, current collector, LVPS, battery, including miscellaneous electrical hardware (breakers, etc.)	75,000
Auxiliary Equipment	Alarm system, wipers, horn, light fixtures, control switches	300,000
Passenger Doors	Side doors and controls.	60,000
HVAC	Cab, floor and overhead heaters, air conditioning and controls.	100,000
Propulsion	Controls, motor and gear unit	56,000
Brakes	Braking controls and equipment, compressor and air supply, master controller	70,000
Couplers & Associated Equipment.	Automatic coupler, drawbars and trainlines	200,000
Monitoring System	Central Monitoring System	80,000
Overall Subway Car with A.C. type Propulsion Equipment - MDBRF - 9,870		

5 CARBODY STRUCTURE

Although stainless steel is historically the preferred material of choice for most North American subway car transit operators, the TTC have steadfastly retained aluminium for their carbody superstructures. Thus visually the cars, which retain brushed aluminium side skins and side doors, have remained basically unchanged since the class M- cars were first introduced in 1962.

Essentially, a semi-monocoque aluminium superstructure is attached to a relatively stiff underframe (ref. 6). As demonstrated in a specified Type Test, the structural shell can withstand a 210 kN (47,250 lb) vertical load applied simultaneously with an 890 kN (200,000 lb) end load without permanent deformation. The strength of attachment of the collision posts to the end weldment is 890 kN (200,000 lb) of ultimate shear at the level of the end sill. The attachment of underfloor and roof-mounted equipment can withstand forces associated with the following decelerations of the equipment mass:
- +/-2 g vertical
- +/-3 g lateral (plus equipment tare weight)
- +/-4 g longitudinal (plus equipment tare weight)

Steel cross channels carry the equipment loads into the side sills. As is customary in North America, the equipment boxes are either supported by attachment bolts in safety or in shear.

The centre of the carbody is built with a slight positive camber with respect to the carbody bolsters at the tare condition. The vertical carbody loads are transferred into the trucks through sliding side bearers on the underside of the carbody bolsters. Loads in the horizontal plane are transferred to and from the truck through a centre pin, which also permits the trucks to be raised with the carbody. Horizontal loads from the coupler are transferred into the draft sill through a ball anchor.

Since the class H5, the end frames of all new subway cars have always been built with vertical collision posts tied together transversely at side plate level by anti-telescoping beams. These collision posts are attached to the end weldments of the underframe by bell-crank shaped anchors. Anti-climbers are welded to the end sills of the end underframe weldments.

Whereas the existing vehicles have 1,143 mm (forty-five inches) door apertures, those on the T1 vehicle are 1,524 mm (sixty inches) wide to increase passenger flows at important interchanges. Thanks to the severity of the Canadian winter, side doors have to retract into door pockets. The wider doors have dictated longer door pockets and thus the window openings on the T1 vehicle are smaller than those on earlier cars. By careful attention to detail, the "porthole effect" has been avoided.

Although the T1 car has fewer seats than does the H6 car (66 v. 76), a greater number of standees can be accommodated. With the existing design of car, the wider side door apertures would have dictated longer nested steel inserts. The total length of all such inserts would thus have increased to at least two-thirds the length of the aluminium side sill. With the greater passenger loads, the greater projected equipment weights and the greater propensity for future corrosion problems, the aluminium side sill was replaced by a Z-shaped steel side sill.

The real benefit in having a corrosion-resistant, low-alloy, high tensile steel underframe is that it will readily exceed the specified fifteen minute exposure to an ASTM E119 test as demonstrated earlier on an almost identical underframe built earlier for the Massachusetts Bay Transportation Authority (MBTA).

One small departure from previous practise has been the deletion of the roof gutter from the extruded side plate. Disposing of the water which collects in the seventy-five foot long gutters has never been entirely satisfactory, staining of the vehicle ends usually being the result. Mini gutters above each of the four doorways per side are now provided; accordingly, there is far less water run-off and any subsequent staining of the side skins is concentrated into areas which are easily accessible. As the vehicles are given a mild phosphoric acid wash every six to seven days, staining has been eliminated.

6 CAR INTERIOR

The interior side, ceiling and end materials have been upgraded from earlier cars to meet the U.S.A.'s Federal Railroad Administration Smoke and Flammability standards.

The interior seating arrangement, developed from the T1P car, is a combination of transverse and longitudinal seating. As noted earlier, the wider doorways have resulted in the T1 having ten less seats than the H6 car. The centre line of the transverse seating is not coincident with the centres of the windows. The seat fabric is both fire and vandal resistant.

Fig. 2. Car Interior.

A new feature to the subway car is a three-place flip-up seat assembly which retains a wheelchair when the seat is in the up position. A passenger assistance alarm tape switch and the electro-mechanical lock control switch and release lever on the underside of the seat are readily accessible to the wheelchair patron.

The car interior colours are burgundy and grey, the TTC's house colours. The Commission's traditional style of courtesy screen has been retained.

As noted elsewhere, the selection of a new all-electric single handle controller has permitted the layout of the cab to be greatly improved. The prominent location of the monitoring terminal unit on the left-hand side of the console will be noted. Such a layout was developed jointly between the union representing the TTC's motormen, the TTC's T1 Programme Management team and Bombardier.

The front console is arranged with the speedometer and air gauge directly in front of the driver. Annunciator lights are clustered on the right-hand side and the monitoring screen on the left-hand side provides both the driver and guard with data about the health of the train's systems.

Fig. 3. Cab Console.

A new car-to-wayside radio-based communication system will be installed on the T1 cars once the system-wide wayside antenna (leaky RF cable) have been fully installed. Not only will the new radio system offer improved communications, it will also enable data about the vehicle system status to be transmitted to Transit Control.

The motorman sits with his back facing a large electrical equipment cabinet. To realize Mean-Time-To-Replace (MTTR) goals, access to the internal equipment can be gained from three sides of the cabinet:

- By-pass and cut-out switches are accessible to motormen and travelling line mechanics from the front of the cabinet. Grouped together on a separate panel are five RS232 connectors to which portable test equipment may be connected to interrogate the car`s microprocessor-controlled systems (Car Monitoring Unit, Propulsion System, Static Invertor, Brake System and the HVAC system).
- The fuse, circuit breakers and door control relay panels are accessible from the side
- The rear door of the electrical equipment rack provides access to the friction brake electronic control unit, the car monitoring unit and the cab control unit for specialized depot staff

A point of interest to note about the cab is that the side sliding window on the right-hand side of the car, although somewhat smaller than its predecessors on earlier cars because of the extended door pocket, still has the same viewing area. The door control panels are in their customary position below the side sliding window. The width of the cab door is such that it can latch onto a transverse seat opposite the cab thus permitting the guard to perform his duties on both sides of the vehicle at a point somewhere in the middle of the consist.

Both windshields at the cab end of the car have impact-resistant glass.

Up to and including the class H6, Toronto`s subway car fleet has been equipped with one rotary door operators per doorway, either electric or pneumatic, with a suitable overhead conjugating bar. Following their successful experience on the T1P car, however, the Commission selected Vapor long-stroke pneumatic door operators for the T1 car with the two individual door leaves linked by a drive belt. Experience gained from the T1P car showed that it was best to locate the associated control manifold for each operator behind an adjacent hinged panel to one side of the door aperture for ease of access.

The side doors are top-hung and are guided by a thrust strip. Before the doors close, warning chimes sound.

The aluminium door threshold extrusions are heated along their entire length by replaceable thermostatically-controlled heating elements energized from the 120/208V AC system. (The TTC still park as many vehicles as possible in the tunnels overnight during severe weather conditions).

7 TRUCKS

The UTDC-designed T1P fabricated truck was subsequently selected for the T1 (and Ankara) contracts. However, major structural and geometrical changes had to be made to the truck frame to permit their operation on vehicles with higher projected gross weights, notably:

- chord thicknesses 50% greater in those critical areas predicted by FE analysis
- ground edges of the chords at those areas with high stress concentrations
- larger transition radii in those plates with high stress concentrations
- welds transferred to lower stressed regions of the frame

The fatigue load spectrum was determined by measuring dynamic motor accelerations, truck frame accelerations and truck frame stresses at critical regions of the T1P frame whilst in service. The all-important transom to side frame connection of the Bombardier- designed truck now has a 50% higher fatigue strength than does the similar connection on the earlier UTDC-designed T1P truck.

Fig. 4. Truck.

Higher projected vehicle loads dictated the use of larger class "E" inboard tapered roller journal bearings which, in turn, dictated a redesign of TTC`s existing 711 mm (28 in) diameter class "C" wheel plate. Side-mounted Wabco BFC wedge-type tread brake units (TBU`s) have replaced back-mounted TBU`s so as to meet Mean-Time-to-Repair (MTTR) goals. The TTC`s traditional cable-operated handbrake has been retained.

Whilst the TTC`s standard current collector has been retained, the open ribbon fuses have (finally) been replaced by enclosed high rupturing capacity fuses.
The classic, longitudinally-mounted motor arrangement (derived from PCC practise) is retained.

The torque from the new, self-ventilated AC motors is transferred into the wheelsets through cardan shafts and proven hypoid, right-angled boxes with a 63:9 ratio from ZF Hurth. Such an arrangement permits a motor to be changed out by a two man crew from a pit road within three (3) hours.

A dual isolated ground brush box is mounted on the inboard side of the T1 gearbox as opposed to the single brush box on the H6 vehicle. One ground brush handles return traction currents, the other fault currents. The H6 and T1 gearboxes are fully interchangeable and, as described elsewhere, the speed sensors are now mounted on one of the end bells of the motor frame whereas on the class H6 they sensed the passage of the crown wheel of the gearbox.

The secondary suspension employs double convoluted air bags with secondary roll control provided by a classic three-point levelling valve system.

Very high longitudinal forces were unexpectedly measured in the inboard-mounted drag links transmitting horizontal loads between the truck frame and bolster of the T1P truck. Designing welded anchorages with a suitable fatigue life to be able to resist such forces or relocating the drag links outboard where lower forces could be expected proved to be difficult. Ever mindful of the schedule and wary of both the technical risk and reliability requirements, Bombardier adopted an updated version of the classic cheek plates design whereby traction and braking loads to and from the carbody are transferred directly into the side frames.

The weldment on the end of each side frame accepts the Commission's recovery dolly, a feature which once installed permits a subway car to return to depot, albeit under greatly reduced speed, should a gearbox, drive shaft, axle or bearing fail in service.

An over-rotation sensor on the underside of the carbody bolster looks at a curved land atop of the truck bolster. Should a truck split a switch, the sensor will rupture the emergency brake loop circuit causing an emergency brake application. Another component which will also cause an emergency brake application is the trip switch. While thought was given to exhausting a small reservoir with the resulting pressure detected by a pressure switch, in the end, a service-proven electric trip switch used by the Chicago Rapid Transit Authority was eventually adopted.

And finally, one will note that grease guards are employed, it being the TTC's practise to grease the guard rails adjacent to the low rail.

8 COUPLERS

The automatic, remote-controlled mechanical couplers and associated drumswitch at the cab end of the car, although similar to the equipment on the earlier class "H" cars, have a few important differences, notably:
- The brake pipe port on the coupler face has been deleted
- The electric coupler box is now fastened below the face of the mechanical coupler whereas previously such boxes were side-mounted

Thus the T1 couplers are only mechanically compatible with the existing fleet and will only have to be coupled to the earlier cars during emergencies. (An emergency electrical connection is provided for the intercom, bell and buzzer should it be necessary to couple to existing cars during an emergency).

At the non-cab end of the vehicle, semi-permanent drawbars connect the married pair, the first time the TTC have used such a component on such widespread a scale. The electrical connection between the married A- and B-cars is accomplished with electrical cables with plug and socket connections.

As is customary, the draft gear deflection in both the automatic coupler and the semi-permanent drawbar is a nominal 38 mm (1.5 in) under an applied load of 512 kN (115,000 lb). At this magnitude of applied buff load, shear bolts within the draft gear fail and the coupler head moves back permitting anti-climbers on opposing cars to engage. On the T1 vehicles, "Ringfeder" springs have supplanted the rubber elements used up to date in the draft gear as the TTC`s experience is that the latter are more prone to loosing preload which then results in unacceptable wear of mechanical coupler components.

With this North American style of coupler, the coupler rotates about the ball anchor and is supported by a spring carrier and radial slide bar.

Of interest to note is that the electric contact pins are trace-heated by the 120/208 V AC system through an ambient air thermostat.

9 BRAKING

9.1 Introduction
Since the introduction of the G-cars, the TTC`s subway cars have either used the Westinghouse Brake and Signal`s "Westcode" seven-step, coded digital electro-pneumatic (E.P.) brake, or latterly, the Wabco North American derivative of this brake, the "Servotrol". Both of these systems are superimposed on the classic brake pipe (BP) whereby the BP pressure can be modulated to produce a service brake application or totally vented to produce an emergency brake application.

Well aware of braking developments on subway vehicles in such cities as Singapore, Hong Kong and London and anxious to reduce the overhaul costs with the associated BP pneumatic valves, the T1 specification permitted Carbuilders to propose either a single pipe system with an electric emergency brake (EB) loop continuity circuit or an "emergency pipe" system.

(The latter scheme permits an emergency brake to be pneumatically propagated from any car on the train but will be electrically assisted should the master controller in the active cab be put into emergency or the deadman position. Such a system was proposed by Bombardier in an endeavour to meet the rigorous reliability requirements of the T1 specification).

In any event, the TTC selected the Wabco RT5 single pipe system with microprocessor control.

Although the electrical control of emergency braking is now fairly universal on modern rapid transit rolling stock outside of North America, this is the first occasion on which such a braking system has been employed on a North America subway car. When ruptured, this circuit produces an irrevocable, maximum load weighed EP brake application on each car of the train; simultaneously, the traction power and electric brake control trainlines are de-energized. It is described more fully in a later section.

The service brake is an energize-to-release, E.P. analogue self-maintaining system providing infinitely variable friction braking.

The electric (dynamic/regenerative) and air brake provide the following system functions:

- Blending of the electric and electro-pneumatic braking
- Preferential selection of the electric brake over the friction brake up to the limit of capacity of the electric brake
- Blending in of friction braking beyond the maximum electric brake limit until the combined brake effort satisfies the brake demand signal
- Loss of electric braking on any one car in the train will result in the total braking effort on that car being provided by the friction brake

On each car, the brake demand signal is load weighed and then modified by the dynamic brake achieved and fade signals.

The PWM control system for blending the electric and EP braking was adopted from the TTC's Articulated Light Rail Vehicles (which featured Brush chopper propulsion and a Knorr single pipe air brake). Out of interest, these vehicles employ electrical control of emergency braking in which solely the B+ feed has to be ruptured.

9.2 Master Controller

An all-electric combined traction and braking master controller with fore-and-aft motion of the handle has been introduced for the first time on a TTC subway car. As noted earlier, this has been of great value in improving the layout of the cab console.

The forward-most position is parallel (i.e., full power) and the position closest to the operator is emergency brake. In addition, detents are provided for full-service brake, coast, inch (i.e., minimum power), series and parallel to mimic settings on earlier cam controlled cars. The deadman function is prevented by keeping the handle twisted against a spring at any position when there is less than 80% of full service brake otherwise the brake system will go into emergency. (When first introduced, the deadman function was prevented by keeping the handle twisted and depressed. As this design gave rise to a documented rash of carpal tunnel syndrome amongst the motormen, it necessarily had to be replaced). The handle position is determined by an absolute, one thousand and twenty-four discrete position optical encoder within the master controller.

The associated reverser is a three-position, handle-operated switch that is mechanically interlocked with the master controller. It has two high-reliability optical switches connected electrically to the encoder/decoder which generate the forward and reverse trainline command functions.

9.3 Friction Brake Electronic Control Unit

The friction brake electronic control unit (FBECU), an encoder/decoder, is the interface between the master controller and the trainlines. It establishes the active cab by reading the forward and reverse trainlines and checking the state of the cab reverser switch. If neither trainline is energized and one of the switches is on, then the FBECU will assume controlling cab status on that car and accordingly the TLREF relay will be energized to establish the controlling car ground reference.

The FBECU generates a battery level 100 Hz PWM rate signal for control of traction and braking. It also receives a 100 Hz PWM electric brake feedback signal from the propulsion logic to determine the level of brake cylinder pressure required.
The FBECU performs internal and external diagnostics and the annunciation for some of the friction brake equipment in the following areas:

- Static diagnostics such as a control test made by maintenance staff from the active cab to troubleshoot operational control problems
- Dynamic diagnostics made when the vehicle operates to determine the health of the friction brake control components

The FBECU communicates with the train monitoring system (TMS) via a dedicated communication link in accordance with a Bombardier serial link communication protocol.

9.4 Pneumatic Equipment
The pneumatic control unit (PCU), a series of valves mounted on a common manifold, responds to commands from the FBECU and provides the requested amount of brake cylinder pressure to the tread brake units (TBU`s) under either service or emergency braking conditions.

The air spring and brake cylinder control pressure transducers which provide control signals to the FBECU are also mounted on the PCU valve-manifold.

A two cylinder, two-stage reciprocating compressor is provided on each B-car and distributes compressed air through the trainlined main reservoir pipe. The compressor is driven by a 208V AC motor and has a 12.5 L/s (26.5 SCFM) output so as to meet the Commission's 50% compressor duty cycle requirement. A twin tower, continuous duty, automatic cycling air drier is located on the frame of the compressor.

Adjacent to each truck are the wheel slip valves (WSV) controlled by the FBECU which, when energized, dump air from the TBU`s on that truck. In emergency braking, the WSV`s are de-energized. Braking effort is reapplied once the slide is corrected.

9.5 Electric Control of Emergency Braking
Electric control of emergency braking was first introduced on British Rail`s P-E-P train of the early seventies (ref. 6). The system developed by Wabco and Bombardier for the T1 vehicles has a good deal more redundancy incorporated and is most similar to the emergency braking system used on the former UTDC`s Advanced Rapid Transit cars on BC Transit.

The sole function of the Emergency Brake (EB) loop is to propagate an emergency brake command. The EB loop is a vital, battery level loop circuit made up of four trainlines EM1, EM2, EMC1 and EMC2

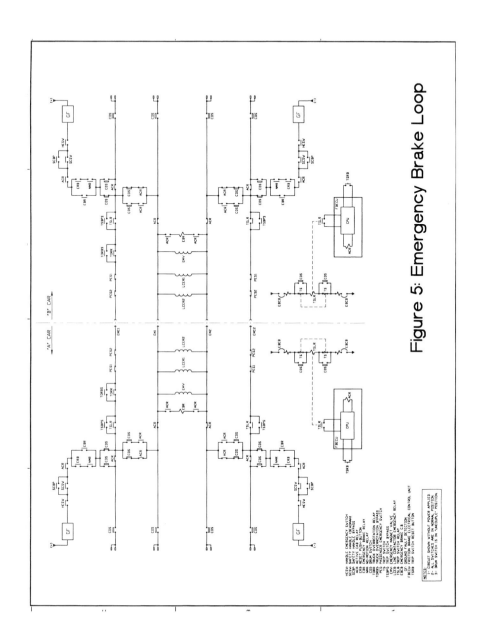

Figure 5: Emergency Brake Loop

The active (controlling) cab supplies B+ (nominal 37.5V) to trainline EMC1 which has all the positive conditioning contacts. At the end of the train, EMC1 loops through the drum switch where it is connected to EM1 which provides the positive voltage to all of the EB loop reading devices.

EM2 provides the negative return for all the EB loop reading devices. EM2 loops through the drumswitch at the end of the train where it is connected to EMC2. Trainline EMC2 has all the negative conditioning contacts and is terminated at the active cab`s B-. Thus all critical functions break both the B+ and B- sides of the emergency trainlines.

The emergency braking control system fails in a fail-safe manner, i.e., when control system elements such as the low voltage power supply fails, the train goes into EMERGENCY.

To protect against chain failure, redundancy has been added. This results in an almost duplicate chain of contacts between the bottom of the group of coils and the power supply return.

Under normal conditions, a group of coils is kept energized. These are:

- The EMV (the R5D transfer valve in the PCU) on each car
- The LCER1 (Line Switch contact in Line Switch Box 1) on each car
- The LCER2 (Line Switch contact in Line Switch Box 2) on each car
- The EBR (Emergency Brake Relay) in the active cab

When any one conditioning contact in the EB loop opens, this group of coils drops out. When the EMV drops out, full supply reservoir pressure is applied to the tread brake units on all cars. When the EBR drops out, the EB loop is unlatched. It can only be re-initialized once the no motion reset relay (NMRR) is energized, i.e. once the train is at rest.

Loss of, or low main reservoir pressure opens a pressure switch which causes the FBECU to de-energize the Active Cab Relay (ACR).

Bombardier and Wabco introduced ground fault detection into the EB loop to prevent unacceptable degradation of the loop. When such a fault is detected, the FBECU de-energizes the ACR and logs the event.

10 PROPULSION

10.1 Introduction
At the time of the bid, Bombardier already had a good deal of experience with AC propulsion systems from GE - USA on the MBTA #3 and the Metropolitan Transportation Authority - New York City Transit (MTA-NYCT) New Technology Test Train R110B cars. The corresponding R110 cars for the MTA-NYCT A Division built by Kawasaki had employed the ADtranz AC propulsion system and this was the system selected for the T1 cars, the first application of such a drive system on the TTC.

As noted earlier, a pair of prototype vehicles was sent to Toronto well in advance of six, pre-production vehicles to determine if, in particular, the new propulsion system could cause unsafe operation of the track signalling, supervisory control and radio communications network.

A separate power circuit drives each truck of the car. Each such circuit has a line switch box, line reactor, invertor, a pair of motors and a set of braking resistors.

The trainline signals from the active master controller, which are input to the central processing unit (CPU) of the propulsion logic control system, specify:

- the direction of travel
- the operating mode
- the type of operation (power or brake)
- the required amount of tractive effort (rate of acceleration or deceleration)

The master CPU then establishes the rate request (mph/s) and the proper power circuit configuration.

A jerk limited torque request is produced once the rate request is combined with the car parameters (car weight as determined from the load weigh signal and wheel size which is set into the logic). This request is then transmitted to the invertor microprocessor controllers which produce the firing signals which control the GTO power invertors.

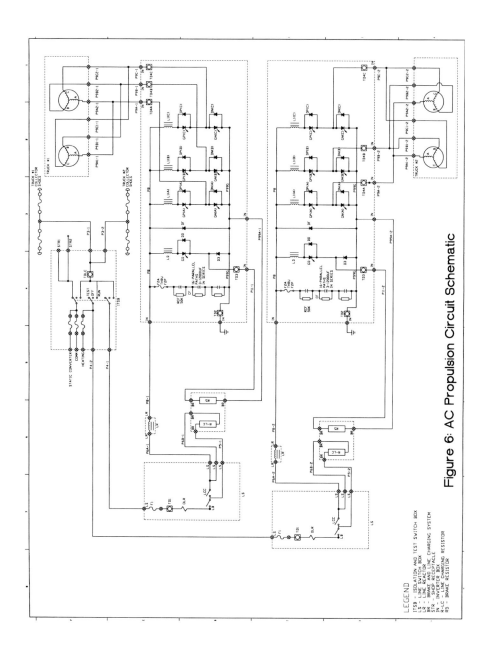

Figure 6: AC Propulsion Circuit Schematic

The principal components of this circuit are described in the following sections.

10.2 GTO Power Invertor

The effective voltage to the motor windings is controlled by the ratio of time each winding is switched to the DC supply voltage versus switched to the ground by the GTO`s. The firing signals ensure that the following conditions are met under all possible operating conditions of the vehicle:

- three phase power is applied to the motor
- the applied power produces motor phase voltages with a fundamental frequency equal to the calculated invertor frequency
- the GTO`s operate within their capabilities (the devices must be allowed sufficient time to turn off and on)
- an essentially constant magnetic flux is maintained across the motor air gap so motor torque is proportional to motor slip (a ratio of about 8.89 volts/Hz is used)

The CPU of the propulsion logic uses the following switching strategies to generate motor voltages that meet these requirements:

- pulse width modulation (PWM) is used from the start-up to about 35 Hz (22 km/h)
- the quasi-six step procedure is used from 35 Hz to 67 Hz (22 km/h to 42 km/h)
- the six step procedure is used above 67 Hz (42 km/h)

In this manner, the invertor provides smooth and precise control of the traction motors over their entire operating envelope.

Each invertor controls a pair of motors on one truck and provides the same level of voltage to each motor. The motor frequency is set at about 7.28 times the invertor frequency to ensure that the magnetic flux across the stator/rotor air gap is fairly constant.

A similar operation controls the generation of voltage in braking. For braking, a negative tractive effort is requested. This corresponds to a negative torque that can be achieved with negative slip. The applied invertor (stator) frequency results in a synchronous speed less than the actual speed of the motor.

10.3 Traction Motor

The ADtranz type 1507A traction motor is a three phase, self-ventilated 104.4 kW AC induction motor with Y-connected form wound stator windings and a squirrel cage rotor. The torque is based on the tractive effort request transmitted by the trainlines.

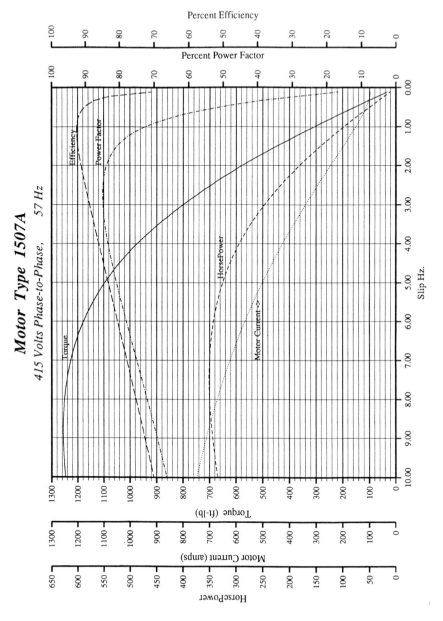

Fig. 7. Operating Characteristics for Type 1507A AC Motor.

The operating curves for the motor at base voltage and frequency which are applicable for operation in both power and braking. The motor weight is 404 kg (890 lb).

10.4 Speed Sensors
Speed sensors are mounted to the frame of the traction motor on the air inlet side. On each truck, one motor has two speed sensors, the other one. They couple inductively with an iron wheel with one hundred and twenty teeth which is shrunk onto the motor shaft just ahead of the bearing cap. The speed signals are critical to the operation of the propulsion system. The signals detect spins, slides, over-speed conditions, zero speed and the motor speed needed for slip and torque calculations. Furthermore, the arc of separation between the two sensors on a single axle indicates the direction of rotation and thus the direction of car movement.

10.5 Dynamic Brake Circuit
The propulsion system provides electric braking for all speeds from 80 km/h to dynamic brake fade-out. Supplemental friction braking is required, however, at both high and low speeds.

The maximum dynamic braking rate for the vehicle is 1.3 m/s^2 (2.9 mph/s) up to 88 km/h. (55 mph) at W1 load. Above 56 km/h (35 mph) at W4 load, the dynamic brake effort tapers.

Friction braking is required for the full service brake rate below about 5 km/h (3 mph). Linear tapering of the electric brake rate is started at about 8 km/h (5 mph) so that the friction brake can be established before electric brake fade-out.
As mentioned earlier, a 100 Hz PWM dynamic brake feedback signal is output to the friction brake equipment to indicate the level of dynamic brake being achieved.

The dynamic brake circuit, comprised of inductor LD, GTO module GD, back diode DD, diode D3 and brake resistor R5, provide a load for the electrical braking energy. This circuit is chopper controlled at a frequency of 200 Hz using the voltage across the filter capacitor CF as a reference for the firing signals.

If the line is not completely receptive, then the voltage across CF rises and when it reaches 870 V, GD is turned on and current is diverted into resistor R5. At the end of the 5 millisecond cycle, GD is turned off and current is allowed to return to the line.

Resistor R5 is sized to supply dynamic brake capability for one complete stop when the receptivity is zero. A software programme estimates the temperature of the resistor tubes and disables dynamic braking if the calculated rise exceeds 800 degrees Celsius.

10.6 Braking and Line Charging Resistors
There are two complete sets of braking and line charging resistors mounted in a common assembly. The line charging resistors are switched into the circuit when power is requested to the traction motors. When the line switch picks up, the line charging resistors are put into the circuit to limit the inrush current to the filter capacitors of the input filter. When these capacitors are charged to within 100 V of full line voltage as measured by the propulsion logic, the line charging resistors are shorted out allowing full line voltage to be applied to the propulsion circuits.

During braking, the line voltage is sensed to determine if the line is receptive for regeneration.

If the line is only partially receptive, the propulsion logic modulates the firing of a GTO so that the remaining portion of the energy is dissipated in the braking resistors. As the line becomes more or less receptive, the logic adjusts the on/off time of the GTO to make up the difference.

10.7 Line Switch Box
The line switch box disconnects the line voltage from the propulsion system under the following conditions:
- when directed by the logic during an overload
- when directed by the logic during low line current condition
- when the overload relay senses an overload condition
- when in the brake mode, as directed by the logic to prevent regeneration during a rail gap or into a dead section
- during truck circuit isolation

The line switch/overload relay is rated for 400 amps rms current and has position sensor feedback to the propulsion logic and low current blowout coils for quick elimination of arcing when there is low current.

11 HEATING, VENTILATING AND AIR CONDITIONING

Ever since the class H5, introduced in 1976, the Commission's newer subway cars have had conventional, split system air conditioning equipment.

Typically, the compressor-condensor unit has to be mounted on the underframe adjacent to one of the trucks (so as that the condensor grille will remain within the vehicle equipment gauge lines). A separate blower-evaporator unit is attached to the roof structure above the ceiling at each end of the car; the depth of these units has always dictated that the interior ceiling height at the ends of the car be slightly reduced. Mixing plenums and filters for the fresh and return air flows are incorporated into each such unit.

The heating, ventilating and air conditioning system (HVAC) provided on the T1 cars by Westcode, U.S.A., is designed to give interior conditions of 25.5 degrees Celsius, 55% relative humidity with an exterior temperature of 35 degrees Celsius dry bulb, 24 degrees Celsius wet bulb and a two hundred passenger load.

Unlike systems on earlier cars, the blowers on the T1 pull the air through the evaporators. Each overhead blower unit delivers 51 cubic metres per minute (1,800 cfm) of conditioned air (20% of fresh air, 80% of re-circulated air) to the duct work in the high ceiling, where it enters into the car through ceiling-mounted diffusers. A diagonal splitter in the duct work ensures that each blower-evaporator unit supplies conditioned air over almost the full length of the car. Conditioned air for the cab is taken from this main duct and discharges into the cab through the ceiling.

Fresh air is drawn into the mixing plenum through flexible ducts on each side of the plenum. The fresh air can be adjusted from 0 to 100% of flow via adjustable fresh air dampers in each of the corresponding (rigid) ducts in the ends of the roof.

The cooling capacity of the system is rated at 45.7 kW (156,000 BTU/hour) using R22 as the working media. The operation of the HVAC system is fully automatic with the operational mode being based on the car air return temperature as detected by sensors mounted in the mixing plenums. The temperature set point for each operational mode are factory set but can be changed by the portable test equipment (PTE).

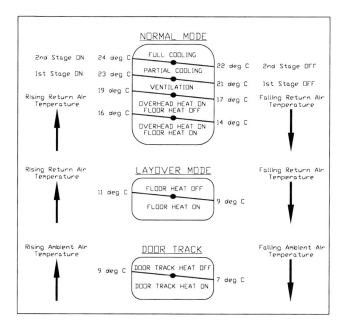

Fig. 8. Temperature Control Chart.

The temperature control unit (TCU) is a microprocessor-based controller which controls all HVAC operations. A portable IBM-compatible PC can interrogate the TCU via the RS232 serial connectors in the cab or on the TCU cabinet on the underframe. Internal operation, contactor operation and system operation are monitored by the processor and fault diagnostics are performed. Within the TCU cabinet and visible from a pit road are a row of LED's which indicate the operational status of the system contactors. A status message is also sent to the car central monitoring unit (CMU) to indicate normal operation or servicing required.

In the partial cooling mode, only one of the two circuits in each evaporator coil is active; in full cooling, both circuits are active. If no cooling is required, then the system will operate in the vent mode.

The semi-hermetic compressor has six cylinders arranged in three banks of two. Two of the three banks are equipped with pressure unloaders to automatically match compressor capacity with loading. Thus the compressor can operate with either one or two banks unloaded, for full capacity, two thirds capacity, or one third capacity. Suitable low and high pressure protection is provided for the compressor. The associated refrigeration control box is positioned on the frame of the compressor-condenser unit and is readily accessible from a pit road.

The compressor is fitted with a crankcase heater and a temperature sensor in the crankcase. The associated controls will not allow the compressor to start until the heater raises the oil temperature sufficiently to remove any liquid refrigerant from the oil and/or to keep the oil viscosity low.

The air-off side of each evaporator contains a single stage of electric resistance heating rated at 7 kW and energized as are the floor heaters from the 600 V supply. The floor heaters are rated at 10 kW.

Now that the cab console no longer has the old-style master controller, a better forced air cab heater has been provided with an adjustable damper at the base of the console.

12 AUXILIARY ELECTRICAL SYSTEMS

12.1 Static Convertor

Both the class H5 and H6 employed motor-alternator sets on each car to supply the air conditioning loads and the separately-excited traction motor fields. The H6 M.A. control sets, as delivered, proved to be mechanically and electrically unreliable and had to be largely rebuilt. This prompted much interest in the benefits of static convertors.

Accordingly, as mentioned earlier, a pair of NEI Control Systems-supplied AC synchronized static convertors were installed on two of the last class H6 cars. In-service trials proved to be highly successful and these convertors are running to this day. For the T1 order, NEI (who in the meantime had become R-R Industrial Controls) offered static convertors with either AC or DC synchronized links. In any event, the DC link system was subsequently selected.

The microprocessor- controlled static convertor has two-stages of power conversion and communicates on a full duplex link with the train monitoring system.
The first stage converts the 600 V DC line voltage into an intermediate 400 V DC voltage. The second stage inverts this intermediate voltage to 208 V, three phase AC.

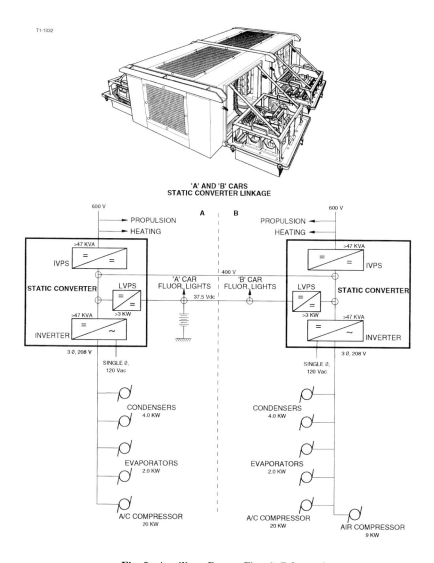

Fig. 9. **Auxiliary Power Circuit Schematic.**

The intermediate voltage on the married pair is interconnected via a DC busbar across the semi-permanent drawbar. The first stage of conversion is so sized that the intermediate voltage power system (IVPS) on one car can provide continuous operation of all auxiliary loads on both cars should one IVPS not be operating as, for example, when the vehicles pass over rail gaps.

The AC is supplied at a frequency of 60 Hz 208 V phase-to-phase, star connected with the centre grounded to the carbody. The resulting phase-to-ground voltage is 120 V. The frequency of the output is within +/- 0.1% or better.

Galvanic isolation is employed. The invertor controls and power electronics are housed in the sealed undercar container with external heat sinks as shown in figure 9.
On the side of the static convertor visible from a pit road are two viewing panels through which one may view the LED's which indicate the operational status of the system contactors.

Maintenance personnel can, by using portable test equipment, access tagged events and faults and historical and pre/post fault information of the operation of the static convertor.

12.2 Low Voltage Power Supply
The IVPS on each car powers a low voltage power system (LVPS) which is connected in parallel to the LVPS on the opposing car. Each LVPS can provide sufficient 37.5 V DC power for all the low voltage loads on both cars of the married pair. As will be noted from the schematic, these loads include battery charging, door controls, heating and air conditioning controls, emergency lighting, marker lights, headlights, operator to guard signal system, electric horn, communication system power and trainline controls.

12.3 Storage Battery
A nickel-cadmium storage battery of 287 AH capacity at the five hour rate of discharge is provided on each A-car. This capacity enables those loads normally powered by the LVPS to operate for at least one hour should the LVPS be inoperative.

In addition to permitting the low voltage supply bus to float charge the battery, the associated battery contactor also prevents deep discharge of the battery when the static invertor is not functioning. Below 26 V, the contactor opens and disconnects the battery from its loads.

12.4 System Polarities
The 600 V DC traction supply power system has its negative pole grounded to the running rails and its positive pole connected to the third rail. The low voltage system on the vehicle has its negative pole grounded to the carbody.

13 TRAIN MONITORING SYSTEM

The Train Monitoring System (TMS), a new feature on the TTC subway cars, is an on-board system with monitoring, diagnostic and data logging capabilities which:

- monitors and logs vehicle sub-system status critical to the operation of trains in revenue service and provides cab annunciation of train status for the operator, guard and (travelling) line mechanics
- provides comprehensive monitoring, diagnostic and data logging capabilities on the five microprocessor-controlled vehicle sub-systems for maintenance personnel
  Bombardier and Vapor North America, the supplier of the TMS, had developed this system on the MTA NYCT R110B cars.
  The Car Monitoring Unit (CMU) on each vehicle acquires, processes and records local

car information and transmits (and receives) this information to (and from) the other CMU`s throughout the train through a high speed network.
The TMS is thus comprised of all the linked CMU`s in the consist. It is a non-critical system, one which is not essential to train operation.
The CMU draws information from three sources:

- the microprocessor controls for the HVAC, static convertor, propulsion and brake systems through a serial link
- discrete 37.5 V dc inputs from the other monitored systems (doors, battery, air compressor, etc.)
- the car identification through 5 V TTL inputs

A Monitoring Terminal Unit (MTU), mounted on the console of each cab as shown in Figure 3, is the CMU display unit.

The base screen allows selection of any subsystem screen or event review screen and displays the train`s configuration and status (such as side doors and handbrake) to the motorman or guard. Several more elaborate screens which display equipment status as well as fault diagnosis can be accessed with a maintenance key.

As noted earlier, a cluster of five RS232 terminals in the rear cab wall permits portable test equipment (PTE) to communicate directly with each of the five microprocessor-controlled systems.

The PTE can perform the following functions on the TMS:

- To collect real-time data from the CMU logs
- To update the resident CMU or MTU software
- To modify the time and date on any one CMU, which then broadcasts the modification to all other CMU`s and to all five intelligent systems

The PTE is an IBM-compatible, laptop PC running under WINDOWS 3.1.

The PTE`s can also be connected to the logic controls of the intelligent systems on the underframe of the vehicle.

14 SUMMARY

This paper has described the careful engineering development behind the modern TTC subway car which has culminated in the Bombardier-built T1 car.

The T1 car still retains a few features which have served the TTC well over the years.

In particular, the old rivetted style of construction - a method of construction seemingly in a time warp to European eyes, has enabled the Commission to repair those cars with corroded side sills, those which have struck barn doors and side-glanced one another, in house, rather than be put at the mercy of a Carbuilder. And there again, a utilitarian, rivetted structure is perhaps quite suitable for a vehicle which spends much of its revenue-earning life in a tunnel and in a city where taking the subway is part of the fabric of everyday urban life.

As one might expect, North American suppliers of traction equipment are well represented on the T1 car as Bombardier have forged good technical and commercial ties with these suppliers on other similar transit car delivery programmes here in North America. These suppliers bring a wealth of knowledge about common problems from other transit properties which complements the knowledge gained by the Commission from its close technical ties with these properties. However, European suppliers who in the past have supplied good, dependable equipment to the TTC and other North American transit market are represented.

Already the T1 car is making an impact at TTC Wilson Depot, where the T1 fleet is being maintained: maintenance personnel who once checked and replaced traction motor brushes have been transferred to other duties.

The T1 cars have been fire-hardened in accordance with the latest US Fire and Smoke Emission standards employing materials and design techniques which have been employed on US transit cars for a number of years. With the increasing complexity of the on-board systems, a fault monitoring system has finally been provided for the motormen, line mechanics and maintenance personnel. And anticipating future needs, the vehicle has provision and suitable restraints for a wheelchair.

For the T1, a technically sophisticated vehicle, to have attained such an overall reliability comparatively early in the delivery programme is most noteworthy. Noteworthy too, is that the TTC, a most discerning transit authority, have recently exercised an option for a further one hundred and fifty-six T1 subway cars.

ACKNOWLEDGEMENTS

I wish to thank Bombardier Transportation Equipment Group and the Toronto Transit Commission for permission to publish this paper. I would especially like to thank my colleagues K.St.Pierre, B.Meadwell, and G. Chony for their valuable assistance in preparing this paper.

My personal thanks must also be extended to fellow Professional Engineers James A. Wood and Paul E. Jamieson of Wabco, South Carolina, Brian Dugdale of R-R Industrial Controls and especially Brian Pawson and Phil Meacock of the TTC for their written technical contributions and encouragement.

References
1. "Traction Motor Experience on Toronto Subway" W.G.Jowett
   (Proceedings 1981 Volume 195 No.33)
2. "Subway Car - seventy-five foot aluminium class M- and H-cars (Camshaft Controls).
   (Internal TTC Public Information Pamphlet August 1991).
3. "A Comparison of Electro-Mechanical and Chopper Propulsion Control Systems on
   TTC Rapid Transit Cars" I.G.Hendry
   (American Public Transit Association, October 1981)
4. "Subway Car - seventy-five foot aluminium class H-cars (Chopper Controls).
   (Internal TTC Public Information Pamphlet August 1991).

5. "Concepts, techniques and experience in the idealization of carbody structures for finite element analysis" K.Wong and A.Firman (IMechE Conference Publication C299/85).
6. "The BR High Density Prototype Electric Multiple Unit Stock" D.Ball (Railway Engineering Journal January 1972).

APPENDIX 1

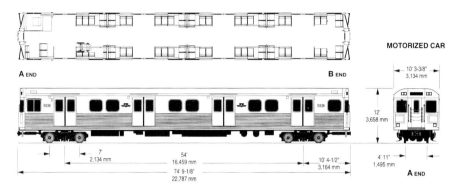

General Data

Type of vehicle	T1 Rapid Transit Car
Owner	Toronto Transit Commission
Date of order	December 1992
Quantity ordered	216 cars (108 power cab cars type A & 108 power cab cars type B)
Train consist	Up to 3 married pairs

Technical Characteristics

- Nominal line voltage 600 Vdc
- Auxiliary voltage 208/120 Vac, 3 ph, 60 Hz
- Low voltage 37.5 Vdc
- 4 induction traction motors (140 hp each - continuous rating)
- 2 traction inverters (GTO)
- Aluminum carbody material
- Two fabricated trucks per car, with inboard bearings
- Pneumatic, regenerative and rheostatic braking system, computer controlled
- Electrical emergency brake loop
- Rubber chevron primary suspension
- Air bag secondary suspension
- Forced convection overhead heaters & convection floor heaters
- Air-conditioning capacity of 13 tons
- Automatic coupling system (at cab end)
- 8 double-sliding side doors
- 1 single-sliding door at each end
- On-board computer controlled diagnostic monitoring system

Dimensions and Weight

	Metric	Imperial
Length over anti-climbers	22,698 mm	74' 5-5/8"
Length over coupler faces	22,787 mm	74' 9-1/8"
Width over side sheets	3,134 mm	10' 3-3/8"
Height (rail to roof)	3,658 mm	12'
Height (rail to floor)	1,105 mm	3' 7-1/2"
Doorway width (side - clear opening)	1,524 mm	5'
Doorway height (side)	1,930 mm	6' 4"
Doorway height (end)	1,981 mm	6' 6"
Floor to high ceiling height	711 mm	2' 4"
Floor to low ceiling height	1,981 mm	6' 6"
Wheel diameter	711 mm	28"
Truck wheel base	2,082 mm	82"
Truck center distance	16,459 mm	54'
Track gauge	1,495 mm	4' 11"
Car weight (empty)		
- Car type A	32,761 kg	72,225 lb
- Car type B	32,384 kg	71,390 lb

Performance and Capacity

	Metric	Imperial
Maximum design speed	88 km/h	55 mph
Maximum operating speed	80 km/h	50 mph
Acceleration rate (standard)	0.94 m/s^2	2.1 mphps
Acceleration rate (high rate)	1.12 m/s^2	2.5 mphps
Braking rate (service)	1.30 m/s^2	2.9 mphps
Braking rate (emergency)	1.39 m/s^2	3.1 mphps
Buff load	892 kN	200,000 lb
Passengers per car (seated)	66	
Passengers per car (standing)	184	
Wheelchair location	1	

APPENDIX 2

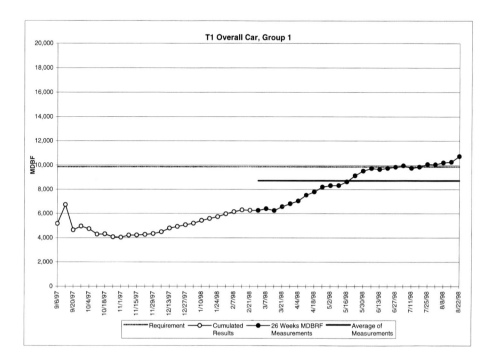

© With Author 1998

C552/043/98

The key to successful, cost-effective, independent safety audit

S BEECH MIMechE
Frazer-Nash Consultancy Limited, Dorking, UK

Synopsis: Following the adoption of the Safety Case approach to Safety Management by a number of industries, the role of Independent Safety Auditor has taken on a renewed importance. Through undertaking the role of, or working with, the Independent Safety Auditor on a variety of projects Frazer-Nash Consultancy Limited has identified a number of key elements which enable this role to be carried out in a manner that is cost effective and of benefit to the whole project. It is the intention of this paper to present these key elements, taken from a variety of industries, for general discussion against the background of the rail industry.

1. INTRODUCTION

Within recent years, industries such as nuclear, defence, shipping and rail have taken on the Safety Case approach as the preferred method of managing safety within their operating activities. An important element within this approach, required by each industry, is that of the Independent Safety Auditor (ISA), whose role is now clearly defined within the regulations and standards relating to each industry. In parallel with the requirement for an ISA, commercial pressures continue to increase, with ever higher targets of efficiency and cost effectiveness being placed on the project. Against this background the cost of an ISA that does not provide any significant benefit to the project, but is in place purely to meet a Safety Management System (SMS) requirement, can be very unwelcome.

It is considered that through a change in the way the ISA relates to the project, the benefits that the ISA can offer to the project can be significantly increased. The key elements of this change are high level, affecting the definition of the ISA's requirements and relationships, and have been observed in a number of projects. It is intended a discussion of these key elements within this paper will be of benefit to those considering the role of the ISA within future projects, assisting in the successful and cost effective application of the ISA to those projects.

This paper will initially consider the requirement for and the aim of an ISA, before the key issues of independence, proactivity, ownership and cost of the ISA are considered.

2. THE REQUIREMENT

The requirement for an ISA is defined within the safety management regulations or standards applicable to each industry. Within the rail industry the requirement for independent audit is detailed in the Railways (Safety Case) Regulations (1). Outside the rail industry, within defence projects for example, a very similar requirement is made within Defence Standard 00-56 (2). This requirement is drawn into each of the three services through a group of Joint Service Publications (3,4,5) which lay down the requirements for the establishment and maintenance of an SMS.

Within these rail and defence industry standards and regulations, and within other industries, the requirement of the ISA is basically common, i.e. independent audit of key safety management activities and of the SMS by a party not directly involved in the design. It is concluded that the similarity in requirements and management processes allows lessons learnt within one industry to be considered within others.

3. INDEPENDENT SAFETY ASSESSMENT

Before considering the role of the ISA further, clarification is required to avoid confusion with the role of Independent Safety Assessor (ISAS). The ISAS role is performed at a lower project level in support of specific design activities with safety implications. This role can be likened to that of the Independent Peer Review in the Nuclear Industry, or that of Design Scrutiny within the rail industry. The assessment takes the form of a detailed design review with the intention of confirming the outcome of the individual design activity. This activity is carried out as part of the design process and is separate from the ISA.

4. WHAT IS THE AIM OF THE ISA?

When starting a new project, which requires the involvement of an ISA, two simple but very important questions must be asked: What is the aim of the ISA? How does this aim fit in with the project?

In order to achieve the successful application of the ISA within a project, it is extremely important to be clear what the aim of the ISA will be. Experience within most industries has shown that without clear definition, or with an aim that does not relate well to the project, it is very easy for the ISA to settle into an auditing role that is distant from the project activities. This may result in a loss of contact between the ISA and the project, and a view being taken by the project personnel that the ISA's presence merely hinders project progress. Previous projects have attempted to define the role of the ISA at one of two levels, either at a low level relating to the audit activities only, or at a high level relating to the overall project aims.

Consider first the role of an ISA whose aims have been related to the audit activities only. This role will concentrate on the task requirements of the ISA, i.e. to audit, and will also concentrate on the maintenance of independence from the project. This approach will ensure that the required audits are carried out, however, these audits could be conducted in a very mechanistic manner, and a manner distant from the project. The interaction with the project is characterised by the generation of a list of actions, often some days after the audit, which may appear to have little relevance to the ongoing design activities. The resolution of such actions may take up a large amount of time with little benefit to the project. This, in turn, may have a destructive effect on the overall project aims, diverting time and effort away from the core design activity. It is found that this approach rarely succeeds in providing a benefit to the project, beyond the basic audit function, and is not perceived as providing benefit to the design team.

The second approach was, therefore, developed to define the ISA's aims with respect to the project requirements specified by the end customer. The most significant effect of a definition at this project level is that the ISA will share the project aim of achieving a safe design, with the design team. In this way the ISA is able to integrate more effectively with the project design team through this common aim. It is considered that such an approach re-directs the efforts of the ISA to provide a more project orientated role, thereby generating more positive input to the project and encouraging the ISA to operate in a more proactive manner. It is also intended that the design team will view the ISA in a more constructive manner, now seeing the role as relevant to the ongoing design process. It is considered that this second approach, if used at the very beginning of any project, would present a firm foundation for a more positive and successful use of the ISA.

The clear challenge for this second approach to ISA is to maintain the mandated level of independence while at the same time yielding the project benefits described.

5. HOW INDEPENDENT SHOULD THE ISA BE?

In the extreme of independence the ISA is provided with rigid lines of communication to the project, and provides the basic audit function required of the role. Such a level of independence is typical of an ISA whose role has not been defined at the project level, as discussed above. The ISA will audit safety activities to a strict timetable, on the basis that these activities are compliant with a specified plan. It will be difficult for the ISA to consider whether these activities are of benefit to the project, or whether the ISA output being generated is of benefit to the safety of the design. This level of independence meets the requirements of the ISA, but, the level of benefit provided to the project is low. It should also be noted that complete independence cannot be achieved if the ISA is being funded directly by the project. At the other extreme, the independence of the ISA from the project activities is lost, and this is equally unacceptable.

There must, therefore, be some compromise in the approach which is both acceptable to the project requirements for ISA as specified by the respective industry standards and regulations, and also provides a genuine benefit to the project. In developing a solution to this problem a role termed the "Proactive ISA" was coined. The aims and role of the Proactive ISA are set at a project level, as discussed above. The relationship of the Proactive ISA with the project is defined in the project Safety Management Plan (SMP). This relationship will detail the core independent reporting lines. It is considered that under these arrangements, the ISA forms a well defined part of the project team, and the definition may be used by the ISA to maintain the required independence. The key element concerning independence and the successful application of the ISA is, therefore, the definition of a minimal project structure which allows the ISA to maintain the required level of independence but does not restrict the ISA activities being undertaken.

It should be noted that within all industries, it is ultimately the responsibility of the ISA to maintain independence from the project. This is considered an achievable requirement on the basis that the role is clearly defined and that an experienced and professional organisation undertakes the ISA role.

6. THE PROACTIVE ISA

In developing the role of the Proactive ISA, two significant phases of activity have been identified. These two phases correspond with the setting up of a project SMS, and the running of the project SMS. For each phase, the emphasis of the Proactive ISA activity is different to ensure maximum benefit to the project.

In the first phase, input from the Proactive ISA at the beginning of the SMS planning and implementation will generate significant benefits in identifying which safety activities are required to meet the project requirements, and ultimately in reducing the risk of these safety activities failing to meet the requirements of the project's final approval process. At this stage the Proactive ISA is present as an independent advisor to the project allowing maximum benefit to be gained from the independent professional advice that is available. The most significant benefits can be summarised as follows:

i) A core SMS can be identified and defined by the project at the earliest stage. The Proactive ISA is able to indicate at this early stage whether this SMS will be acceptable, and what additional features may be required.

ii) The early review of the SMS will allow the identification of any tasks surplus to the main requirement of the Safety Case.

iii) Early involvement of a Proactive ISA leads to more effective integration into the project and project team.

In the second phase the Proactive ISA reverts to the auditor role. The proactive approach to be taken now becomes very important, allowing flexibility with the audit timetable and the audit approach taken. This flexibility is required to consider all parts of the SMS, and to account for variations that inevitably occur within a project programme.

In conducting the audits, feedback from the Proactive ISA in a manner most useful to the project is vital. For the more significant actions this requires feedback at the audit, such that corrective action or preparation for the corrective action can be taken immediately.

7. DEFINING THE SCOPE OF THE PROACTIVE ISA

A common difficulty for Proactive ISA activities, is the accurate definition of the scope of these activities. For a successful application, the Proactive ISA requires a high degree of freedom to examine all areas of the project as required. This may lead to extra effort being required in areas of high complexity or particular significance to the overall project. The project may, however, be understandably wary of providing an ISA with a blank cheque. Careful planning of the Proactive ISA activity is required.

In setting up the SMS it is important that the Proactive ISA works with the project to establish Terms of Reference, and to establish what activities will be required. Key activities associated with design reviews, project safety meetings and key acceptance requirements, can easily be identified. Less well defined items may include support to general SMS activities, subcontractor audits, or audits of activities associated with specific project requirements. In these cases, the number of audits can be defined, however, timing cannot. It is therefore vital that the final scope of work leaves room for flexibility of tasks ensuring the Proactive ISA can operate whenever actions are deemed most beneficial.

8. IMPROVING PROJECT COMMUNICATIONS

A key use by the project team of the Proactive ISA is as a means of improving communication within the project and hence assisting in moving the design forward. This improved communication is highlighted in a number of ways:

i) The Proactive ISA is able to highlight safety issues of concern to the project and place pressure on the resolution of these concerns.

ii) The Proactive ISA is able to provide an independent perspective on the project safety activities. This can aid decision making within the project by providing an authoritative view which is less affected by project pressures.

iii) A critical part of any Safety Case is its acceptance by the client or the regulator. Through a careful and planned use of the Proactive ISA within the design team, a significant reduction in the risk associated with this acceptance process can be achieved, whilst maintaining their independent role within the project. With the Proactive ISA's review and contribution to the SMS and endorsement of the use of this SMS through the conducting of audits, two key elements within the SMS are demonstrated to the Safety Case acceptance Authority. This provides confidence to the authority and significantly reduces any risk associated with failure of the acceptance process.

This improvement in flow of information can result in significant reductions in design team time required to handle ISA related issues, and a reduction in risk of the overall acceptance process.

9. MONITORING PERFORMANCE OF THE PROACTIVE ISA

Projects will require the ISA's activities to be monitored for a number of reasons. These reasons may include the monitoring of project progress, or as a basis for the ISA payment scheme. The measurements by which the ISA can be monitored are, however, far from obvious. An early solution was suggested by monitoring the number of comments made or actions raised. However, these numbers will actually bare little relation to the effectiveness of the ISA or the project. A large number of actions may be the result of a number of situations e.g. a poor ISA raising many actions on a good project to fill a report or a good ISA raising many justifiable actions on a poor project.

Similarly, monitoring the ISA by the number of audits carried out may encourage the production of audits on time for the benefit of the ISA, but not the project. Ultimately, it is considered that there is no way of monitoring ISA activities in isolation, as each may encourage ISA activities for the wrong reasons. As discussed above, it is considered that the most effective Proactive ISA is one where the role is defined at the project level. In the same manner, the goals and monitoring of the Proactive ISA have to be carried out at a project level. Like the project design team, the Proactive ISA can, therefore, only be monitored against key design dates and ultimately the acceptance of the Safety Case by the design authority. It is important that the monitoring of any other features of the ISA role are carried out with care, such that the flexibility and actions of the Proactive ISA are not restricted, and such that the role of the ISA can be constantly developed for the benefit of the project.

10. CONCLUSION

In conclusion, this paper has identified a number of key elements which, it is considered, result in the successful application of an ISA with in a project. The nature of these elements results in a revision to the ISA role, termed the Proactive ISA, which generates significantly more input into the design process. The project will, therefore, employ the Proactive ISA in a significantly more cost effective manner.

It is believed that the key issues presented in this paper, identified from a number of different applications, will be of benefit to those considering the role of the ISA within future rail industry projects.

11. REFERENCES

1) Railway safety cases. Railways (Safety Case) Regulations 1994. Guidance on Regulations. L52. HSE Books.

2) Defence Standard 00-56 (Part 1)/Issue 2. Safety Management Requirements for Defence Systems. Ministry of Defence. 13 December 1996.

3) JSP 430. Ship Safety Management Handbook. Volume 1. Issue 1. Ministry of Defence. January 1996.

4) JSP 454. Instructions for Land Systems Equipment Safety Assurance. Ministry of Defence. June 1997.

5) JSP 318B. Regulation of MoD Aircraft. Ministry of Defence. To be issues in late 1998.

C552/081/98

How MM can add value to the business

M J HUDSON MCIPS
Procurement and Logistics Department, EWS, Nottingham, UK

1. **INTRODUCTION**

Good Morning everyone and welcome to my presentation on how the Procurement and Materials Management function can add value to Business. During the next 20 minutes, I'd like to share with you the somewhat turbulent journey of the privatisation process of British Rail within the Procurement and Materials Management function.

I shall start with an overview of the situation EWS inherited in 1996, then take you through the key developments that the Procurement Department of EWS have made to-date, and finish with an insight into our future plans.

So, to begin with then.

2. **THE BRITISH RAIL LEGACY (SLIDE ONE)**

2.1 **Centralised purchasing/limited supplier choice**
The procurement Department of British Rail was heavily centralised and distant from it's internal customers. Supplier choice for T&RS parts was limited to one distributor with no freedom to source spare part supply on a competitive basis. This naturally restricted market force.

2.2 **Bureaucratic and inflexible procedures**
The nationalised way was overly bureaucratic and very much driven by rules to support British Rail's accounting and auditing policy, and this often overshadowed the reason we all come to work, to satisfy customers.

2.3 Lack of Standardisation

Standardisation of locomotive and wagon design was virtually non existent which meant there was a need to develop procurement strategies for a completely new range of spares and support contracts with every new build. The EWS Class 58 and 60 locomotives being a prime example. The on-cost of managing this range of different vehicle types means increased stock levels, storage costs and administration. Engineers tended to over specify components in isolation without consulting Procurement making it difficult for the supply chain to meet requirements and subsequently poor availability was common place

2.4 Insufficient degree of focus on COST

The concept of total cost was never fully explored within BR's procurement structure. There was a degree of focus upon purchase price within the framework of budget thresholds, but many of the other elements of cost were overlooked such as value for money, reliability, standardisation and over specification.

2.5 Too remote from the end-customer

The end-customer of the organisation was invariably remote from minds of decision makers in Procurement which had the difficult task of satisfying a whole spectrum of internal customers with different needs. To a large extent therefore, the activities of Procurement were not end-customer driven.

2.6 Too remote from the supplier

Arms length and to some extent adversarial relations with suppliers were commonplace within British Rail. Regular contract renewal via competitive tender and change of suppliers was the norm in many product / service areas. The buying power of the industry was not put to best use, and the idea of long term partnership based contracts was kept very low key.

2.7 Didn't capitalise fully on supply chain innovation

The direction of BR's procurement function did not fully capitalise on the potential of the supply chain to add real value to the business, many good ideas within the supply base (to simplify specifications for example) being held back through lack of encouragement to make the change (the culture being not to challenge the status-quo), resulting in inherent un-necessary elements of cost being present in many products, components and services.

3. PRIVATISATION: THE OPPORTUNITIES FOR PROCUREMENT (SLIDE TWO)

Now I'd like to overview some of the key opportunities that were available to EWS at the point of privatisation.

3.1 Creation of an independent Procurement Department

Unlike the franchised train operating companies, EWS made a conscious decision by nature of status of the business to create a new independent, Procurement Department eliminating the need to outsource any purchasing activity. This Department managed from day 1 a complete portfolio of Goods and Services including T&RS spares and major overhauls, civils, infrastructure, plant and equipment, investment schemes, consultancy, through to general purpose items such as cleaners, stationery, clothing, hotels and car hire.

A team were brought together in the Autumn of 1996 combining the best internal Railway based procurement skills within Trainload Freight and Rail Express Systems, and private sector companies.

3.2 Immediate price savings through resourcing
Resourcing a selected range of strategic / high turnover consumable and repairable T&RS items from previous BRB suppliers to new sources had potentially high purchase price savings and tremendous opportunities to incentivise the supply base and improve service.

3.3 Rationalisation of procedures, policies, Terms and Conditions
The opportunity presented itself to review procedures and policies and simplify them, including Terms and Conditions of Contracts and to issue a new Procurement manual.

3.4 Standardisation/variety reduction through investment
The new investment injected from the consortium (led by Wisconsin Central Transportation Ltd.) provided an ideal opportunity to begin the long process of standardisation of traction, rolling stock and IT equipment etc.

3.5 Freedom of choice in the world-wide supply base
EWS was exposed immediately to market forces at the input and output end of the organisation and in order to bring costs down sourced equipment globally.

3.6 Closer supplier relations. Partnership based contracts
Tremendous potential was at our fingertips to shorten the supply chain, become closer to our key suppliers and understand each others businesses to form longer term partnership based contracts and to release dormant innovation.

4. EWS PROCUREMENT: KEY ACHIEVEMENTS FROM PRIVATISATION TO DATE (SLIDE 3)

At this point I'll go into some detail on the opportunities EWS capitalised upon within procurement and the key achievements to date:

4.1 Suite of new supply contracts generating savings of (16%)
Resourcing key spend components such as wheelsets, buffers, springs, brake pads generated savings of 16% to the bottom line. The opportunity was also taken to straighten/realign the supply chain during the 1997/8 fiscal year benchmarking using best practices where relevant. It is worth noting at this point that a 1% saving in purchasing costs can have the same effect on a company's profitability as a 10% increase in sales.

4.2 Enhanced customer focus and improved responsiveness
Closer liaison with the marketing team focusing on external customers coupled with a more intimate working relationship with our internal customers has enhanced the customers focus through the organisation and improved response time to customer desires, this being particularly vital in the dynamic environment we are all in today. Technical and commercial skills have been combined to best effect in the supply chain. The status of Procurement and Material Management has been elevated in the minds of our Engineers over the past 2 years

through direct action and achievement of results. Procurement are no longer considered an Ivory Tower, because of our hands on, can do, attitude.

4.3 New procurement IT systems

EWS has invested heavily in new IT systems including a replacement of the mainframe driven BR IMACS system which was too antiquated to meet the future needs of EWS and the year 2000. The replacement called MIDAS, standing for Materials, Inventory Development and Support was successfully introduced in March this year and is on a SAP windows based system. On line visibility of stocks / orders is now available together with reduced transaction times / purchase order costs and tremendous Edi potential for the supply base.

4.4 Investment in new locomotives and wagons

EWS has placed contracts for supply of 250 new Class 66 and 30 Class 67 locomotives from General Motors and a minimum of 2500 new wagons from Thrall Europa for which the majority of components are of an existing, proven design. Once fully operational, there will be great scope for rationalisation of stock and improved logistics. Investment justification was quickly approved and new wagons and locos are here and running in timescales never experienced in the BR days.

4.5/ Launch of major supplier partnership initiative (A.C.T.)

4.6 Development of partnership based supply contracts

EWS launched it's major supply initiative in March this year called ACT, standing for Achievement through Co-operation and Teamwork, as a result of which partnership based supply contracts are emerging bringing many cost saving initiatives.

4.7 Simplification of low value order processes

EWS has through Business Process Re-engineering simplified the way in which low value requirements are processed. EWS currently processes 11200 invoices per period equating to 145600 per year with 55% of them being for less than £250. By introducing Purchasing Cards, the objective is to increase Purchasing and Accounts Payable efficiency whilst reducing administration costs and to give end users the ability to satisfy urgent requirements direct. Procurement specialists are then able to concentrate on strategic issues. EWS has introduced 50 Cards to date and aims to roll this out to 200 users. Interesting to note that the cost of just raising an order is between £35 and £50 which is a direct immediate saving.

I shall now explain in more detail the ACT initiative and the partnership related benefits experienced to date.

5. THE EWS A.C.T. INITIATIVE (ACHIEVEMENT THROUGH CO-OPERATION AND TEAMWORK) (SLIDE 4)

5.1 100 key supplier brought together in March 1998

In March this year around 100 principal suppliers to EWS were invited to attend a seminar at Pride Park in Derby to launch a new initiative entitled "Achievement Through Co-operation and Teamwork".

5.2 **Key objectives: i) Form strategic alliances with key suppliers**
ii) Reduce total cost through supply chain innovation

he primary objective of A.C.T. is to remove any arms length style agreements and incentivise suppliers to think pro-actively and come forth with innovative suggestions to take out cost from the product range without comprising safety and indeed enhancing safety where possible.

6. KEY ASPECTS OF A.C.T. (SLIDE 5)

6.1 Culture change in supply base (shared destiny)
Suppliers are encouraged to share corporate strategies and long term plans with EWS and vice versa to create a shared destiny based relationship realising that to some extent we both rely upon each other to succeed. (WIN/WIN)

6.2 Sharing of ideas and continuous improvement
Ideas to improve quality/service and reduce cost are actively encouraged. Suppliers are incentivised to improve overall service on a risk/reward basis. For example - a longer term contract for investment in facilities or design improvements. Many ideas which have lay dormant for many years have been revitalised.

6.3 Longer term partnership based contracts
Longer term contracts monitored through key performance indicators are being developed to avoid reporting statistics for statistics sake and focusing purely on the quality gap.

6.4 Elevate the status of procurement and materials management
This initiative created a new view on the role of procurement and material management in relation to the company's top level strategy. All activity is now geared towards the EWS strategy which I'll share with you in a moment.

6.5 Supply base more responsive to needs of our customers
Suppliers are becoming more flexible and responsive to the needs of our customers. Systems are now flexible enough to avoid administrative issues over - using common sense where goods are urgently required for example.

6.6 Think outside the box
People at every level are being encouraged to think literally and laterally beyond the traditional scope to come forward with suggestions however small.

7. EWS PARTNERSHIP SUPPLY CONTRACTS: THE BENEFITS SO FAR (SLIDES 6 A&B)

7.1 Combined Engineering Ventures (VA/VE)
Value Analysis and Engineering innovations are being discussed to take cost out of products without compromising safety, or if possible improving levels of safety.

7.2 Specifications challenged by suppliers
Suppliers are now not afraid to challenge specifications and suggest improvements.

7.3 Reverse Engineering of components
New materials to manufacture components are actively encouraged from our Engineers and our Supplier's Engineers.

7.4 Plans for EDi links
EDi links with 3 key suppliers are planned for 1999 to give 2 way visibility of stocks which will enhance planning and reduce lead times.

7.5 Rationalisation of documents, procedures and contracts
Beaurocracy is kept to a minimum. Many forms have been standardised.

7.6 Better understanding of each others business
EWS and its suppliers are beginning to understand the factors behind the demand and behind the product / service that influence the supply relationship.

7.7 Focus on non-price related cost elements
The non price factors effecting all our products are now being identified and addressed.

7.8 Reduction / better utilisation of repairable floats
Floats of repairable components are being reduced and used for spare parts to give an improved return on the capital we employ. This includes reduced repair cycle times.

7.9 Standardisation of component initiatives
Options for standardisation have become more apparent through close working relations with suppliers.

7.10 Streamlined communication
Points of contact have been clearly identified to reduce lines of communication, and improve responsiveness.

7.11 Improved logistics
The costs of logistics i.e. transport, storage, handling are high for most businesses and closer working relations with suppliers bring to light many of the present inefficiencies.

8. EWS PROCUREMENT / MATERIALS MANAGEMENT: THE FUTURE CHALLENGE (SLIDES 7&8)

8.1 Capitalise on world-wide best practise
To further develop in the UK more of the ideas of the Wisconsin group and share best global practice with other key players in the market.

8.2 Cost reduction

To continue to take cost out of products / services to improve EWS's competitive edge without comprising safety, and reduce the company's bottom line costs through EWS led and supplier led innovation.

8.3 Locate / develop world class suppliers

To ultimately only deal with the worlds best supply base and form long term partnerships.

8.4 Maintain / improve customer focus

Everything must continue to be driven by our customers.

8.5 Improved planning and reduce stocks / lead times

Working closer with internal customers to improve planning will reduce the unforeseen variability in our demand. This in-turn will reduce variability in lead times and allow EWS to drive stocks down.

8.6 Introduction of Purchasing Cards

Purchasing cards will be used to reduce the administrative cost of small value orders.

8.7 Work to achieve EWS's company strategy (SLIDE 8)

8.7.1 Productivity
- Reducing staffing levels and equipment requirements to best practice levels
- Aiming for staff number/trains run ratio of between 4 and 5
- Elimination of waste, duplication and unnecessary bureaucracy
- Reducing the need for direct supervision of workforce

8.7.2 Investment
- Investment in the best resources available to do the job:
- People (in the terms of modern packages and training)
- Locomotives
- Wagons
- Equipment

8.7.3 Commitment to Customer Service
- Attitude   - the 'can-do' approach
- Access   - ease of access and one-stop shopping
- Equipment   - exploiting modern technology to make use of EWS simple and straightforward

8.7.4 Expanding the Business

- To reverse the decline in rail freight and indeed triple the business in 10 years by offering customers the services they want at prices they can afford
- Safeguard existing contracts
- Develop new business with existing customers
- Develop business from new customers with:

- flexible services
- one stop shopping
- new services such as intermodal, and less than trainload
- prices pitched in anticipation of emerging cost reductions
- quality and reliability

9. **SUMMARY** **(NO SLIDE)**

Organisations must focus upon all aspects of costs on a continuous basis to be competitive. Procurement and Materials Management has a key role to play in this process as I hope my presentation has demonstrated. We never lose sight of the customer focus, the customers strive for excellence, which must inspire us to do likewise. The challenge ahead is great, but with a positive can-do attitude it will be met.

10. **FINAL SLIDE (SLIDE 9)**

Final thought.

Keeping trains in service and satisfying customers is all about controlling costs. As this slide shows, if the rope snaps, it can be painful. Procurements and Materials, working together as a team with Engineers will ensure the rope remains intact.

THE BRITISH RAIL LEGACY

- Centralised purchasing / limited supplier choice
- Bureaucratic and inflexible procedures
- Lack of standardisation
- Insufficient degree of focus on COST
- Too remote from the end-customer
- Too remote from the supplier
- Didn't capitalise fully on supply chain innovation

PRIVATISATION:
THE OPPORTUNITIES FOR PROCUREMENT AND MATERIALS MANAGEMENT

- Creation of an independent Procurement Department

- Immediate price savings through resourcing

- Rationalisation of procedures, policies, Terms and Conditions

- Standardisation / variety reduction through investment

- Freedom of choice in the world-wide supply base

- Closer supplier relationship. Partnership based contracts

KEY ACHIEVEMENTS FROM PRIVATISATION TO DATE

- Suite of new supply contracts generating savings of £6m (16%)
- Enhanced customer focus and improved responsiveness
- New procurement IT systems
- Investment in new locomotives and wagons
- Launch of major supplier partnership initiative (A.C.T.)
- Development of partnership based supply contracts
- Simplification of low value order processes

THE EWS A.C.T. INITIATIVE (ACHIEVEMENT THROUGH CO-OPERATION AND TEAMWORK)

- 100 key suppliers brought together in March 1998
- Key objectives:
  ◇ Form strategic alliances with key suppliers
  ◇ Reduce total cost through supply chain innovation

KEY ASPECTS OF A.C.T

- Culture change in supply base (shared destiny)
- Sharing of ideas and continuous improvement
- Longer term partnership based contracts
- Elevate the status of Procurement and Materials Management
- Supply base more responsive to the needs of our customers
- Think outside the box

EWS PARTNERSHIP SUPPLY CONTRACTS: KEY BENEFITS SO FAR

- Combined Engineering Ventures (VA/VE)
- Specifications challenged by suppliers
- Reverse Engineering of components
- Plans for EDi links
- Rationalisation of documents, procedures and contracts

EWS PARTNERSHIP SUPPLY CONTRACTS: KEY BENEFITS SO FAR CONT/D...

- Better understanding of each others business
- Focus on non-price related cost elements
- Reduction / better utilisation of repairable floats
- Standardisation of component initiatives
- Streamlined communication
- Improved logistics

EWS PROCUREMENT / MATERIALS MANAGEMENT: THE FUTURE CHALLENGE

- Capitalise on world-wide best practise
- Cost reduction
- Locate / develop world class suppliers
- Maintain / improve customer focus
- Improved planning and reduce stocks / lead times
- Introduction of Purchasing Cards
- Work to achieve EWS's company strategy

EWS STRATEGY

- Productivity
- Investment
- Commitment to Customer Service
- Expanding the Business

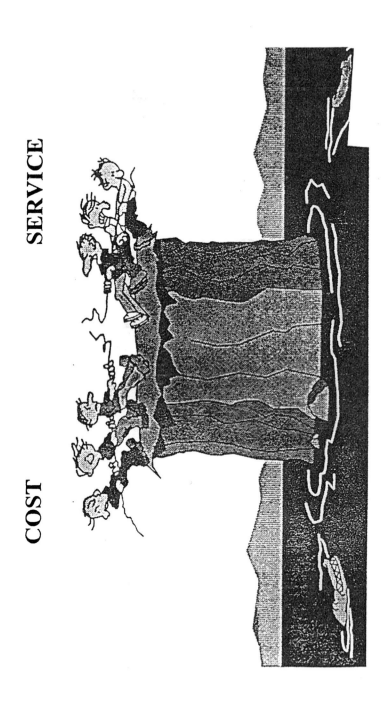

C552/031/98

Whither QRA?

D EDGE MSc, CEng, MIEE, MSaRS
The Engineering Link Limited, Derby, UK

ABSTRACT

In his most recent report on railway safety (1) the then Chief Inspecting Officer of Railways criticised the misuse of quantitative risk assessment (QRA). Without disagreeing in principle, there is a risk that the balance of the Inspector's remarks is interpreted as being against QRA rather than against its abuse. QRA techniques offer great benefits over those methods in use only a decade ago. In particular, the transparency of the QRA-supported decision making process and the rational allocation of resources are reasons for wider application. This paper draws evidence both from general experience of QRA in train engineering as well as some specific tasks which illustrate how QRA allowed better decisions to be made than hitherto, sometimes challenging established practice and received wisdom.

1. BACKGROUND

The railway industry has long prided itself in achieving high levels of safety. It may have been forgotten that much of what is taken for granted was achieved by the Railway Inspectorate against the wishes of railway managements that historically resisted 'unnecessary extravagances' like block signalling, steel bodied stock and continuous automatic brakes.

For most of the working lifetime of engineers in the industry today, the railway was nationalised. By the 1980s unpredictable Government cash restrictions often prevented all but safety investment. Safety expenditure was seen as 'a good thing' although the tools for managing risk were hardly used. Junior engineers without the financial authority to order a box of paperclips could increase the lifecycle costs of a fleet by hundreds of thousands of pounds by adding a few ineffective safety checks to a maintenance schedule. Depot resources were absorbed by such checks to the detriment of work that might have been more effective. A culture existed where safety expenditure was little challenged, but often ineffective.

Privatisation presented the opportunity to link risk and reward and drive out inefficiencies. However, railways are a particularly safe mode of transport (2). How was the market to ensure safety, given that the 'market rate' for transport safety in Britain is about an order of magnitude worse than the railway? The consultation document 'Ensuring Safety on Britain's Railways' (3) proposed a mechanism which sought to demonstrate that the privatised railway would be as safe as before, thus partially constraining economic forces. Organisations which wished to operate infrastructure or trains would be required to show their competence and commitment by means of a railway safety case.

In other industries a safety case would normally be supported by a substantial amount of quantitative risk assessment (QRA). Such assessments would consist of a number of steps.

- both empirical and creative methods to identify hazards, examples being checklists and brainstorming respectively

- reliability engineering methods such as fault tree analysis to quantify the likelihood of hazards

- consequence modelling to quantify the likely impact of each hazard

- identification of effective risk reduction methods and calculation of their cost-effectiveness

- a decision on the chosen design and operating strategy supported by sensitivity analysis and qualitative appraisal.

However, the first railway safety cases were approved with little QRA (4). Emphasis was on describing organisational practices and interfaces. Risks were prioritised using largely arbitrary matrices. This approach is reasonable for maintaining the status quo of an established and relatively safe industry through transition. Further, data was not widely available for carrying out detailed apportionment of risk.

This approach changed with the publication by Railtrack of the Safety Data Compendium (5) and Volume 3J of Railtrack's Railway Safety Case (6). These provided for the first time a convenient means to set risks in context and to apportion overall railway risks consistently. The Railtrack origins added an authority to the data that was possibly never intended. However, it was now possible for QRA practitioners both from the established railway and from diversifying sectors, such as nuclear, to apply QRA in the railway industry to an unprecedented degree.

The result was enthusiasm and confidence that the industry was now in sight of answers to the question 'How safe is safe enough?' and that there was a rational basis for decisions about resource allocations to safety. No longer would priorities be set by the flavour of the month.

2. THE CASE AGAINST QRA DISCUSSED

The honeymoon was soon over. In his 1996/7 report on railway safety (1), the Chief Inspecting Officer of Railways criticised aspects of the use of quantitative risk analysis. The points made are discussed below.

2.1 Major, investment-based decisions

The Inspector states that QRA is only applicable to major, investment-based decisions. Engineers may feel that the rules are being changed now that they are starting to make the economies that privatisation set out to encourage. They understood from the early Health and Safety Executive (HSE) work on the tolerability of risks (7) that QRA was intended to inform decision making and not to replace it. What was less clear was that the approach was only to apply to major investment-based decisions and could not be extrapolated or apportioned.

In fact, it is difficult to support that line of argument. Reliability engineers have long been assisting in the production of safety reports for hazardous installations by using quantified methods such as fault tree analysis to estimate the frequency of major (and minor) incidents. However, historically, in the traction and rolling stock industry, consequence analysis was not practised. All safety-related incidents were lumped into one or two categories. For example a target of 2 million miles between 'hazardous' failures was set for one multiple unit. That loose category could cover everything from failure of the headlight-proving unit to a 15% loss of brake force. This is clearly unsatisfactory, as it does not give priority to high-risk items. Yet only a decade ago, it represented the state of the art in rolling stock procurement.

Now, while it may be the case that the quantified test of reasonable practicability was only intended to be applied to 'major, investment-based decisions', in reality, those decisions were already underpinned by a good deal of low-level, detailed quantitative analysis. The safety of the fleet is inferred from the integrity of the components. Thus if such criteria as '£1M per fatality prevented' could be applied at a high level to the construction of a new fleet, the same criteria could be applied to a vehicle subsystem - '£1000 per millifatality prevented'. There are of course major problems in apportioning risk. The Railway Group aim is less than 1 fatality in fifty million journeys. If 10% are technical failures and 10% of those are T&RS, then what tolerable risk levels should be apportioned to individual components? Axles, one of the most safety critical components of all have only caused one fatal accident in decades. Nonetheless, HSE recommended an investigation that implicitly required QRA following the derailment at Rickerscote that was caused by a broken axle. In any project, a decision needs to be made on when to stop analysis. Apportionment of risk is a useful aid to decision making.

It is not clear what is an 'investment-based decision'. Why should investment-based decisions be any different to ones related to revenue streams? If modifying a fleet at a present cost of £100 000 to achieve a safety target is a legitimate use of QRA, a change to maintenance with a similar present cost should be treated similarly.

2.2 Unjustified faith?

The 1996/7 Annual Report reiterates the position of the 'Tolerability' report that QRA is an aid to decision making, not a substitute for it. The point is well made. It is unusual for both the best and worst case outcomes of a QRA to provide an unambiguous indication of the practicability of a safety measure. In the majority of cases, the decision needs to be justified by a well-argued and thorough qualitative analysis.

The inspector criticises the 'unjustified faith' placed in simple numerical techniques. There is an implication that sensitivity analysis should be used to deal with the considerable uncertainties in the data. A sensitivity analysis should identify all sources of uncertainty. It is, in effect, a risk assessment within a risk assessment. Otherwise, the unwary reader may be lulled into a false sense that the uncertainties have been fully addressed. For example, the recent cost-benefit analysis supporting the proposals for train protection and Mk1 rolling stock (8) appear thorough, but could be optimistic despite their apparent precision (9).

In that analysis, both the best case and worst case values of the impact of crash modifications are given, the latter being approximately 90% of the former. Few QRA practitioners would claim that their analyses justified anything approaching this level of precision. Closer examination shows that these 'best' and 'worst' cases assume amongst other things that the utilisation of Mk1 rolling stock will continue at its present level after the delivery of new stock. However, it is more likely that the stock will be largely relegated to peak only workings, thus reducing the crash benefits. Furthermore, peak hour operation reduces the influence of two contributors to slam door risk - passenger unfamiliarity and alcohol. Thus, although the results are presented as a range of costs of preventing fatalities that appear consistent with industry standards, the actual limits of uncertainty are rather further apart.

2.3 Reasonable practicability and established good practice

HSE are concerned that the application of the technical definition of the phrase 'reasonably practicable' (ie £2.65M per equivalent fatality (10)) was being used to the exclusion of the everyday meaning. It can be shown by reliability engineering methods that certain historical maintenance practices are not reasonably practicable (in the technical sense) and that maintenance intervals could be extended or certain types of maintenance abandoned. HSE argue that, if something has been done previously, it must, in a commonsense way, be reasonably practicable, even if it is very expensive in relation to the risk it mitigates.

This insistence that 'established good practice' is always reasonably practicable; ie that standards should never be reduced, seems reasonable. However, it may have an unfortunate side effect. Existing practice (for reasons spelled out earlier) may not be 'good'; it may be ineffective, expensive, inefficient and even demoralising.

Maintenance is carried out by humans and those humans have ideas about what is effective and what is ineffective; what is possible and what is not possible; what they need to do and what they need not do. The practice of specifying frequent, unfocussed examinations in the hope of preventing component detachment is regrettable as it has the opposite effect to that intended. The integrity of maintenance can suffer when such tasks are specified.

Human factors issues in maintenance have been poorly addressed by the industry. While the Human Factors in Reliability Group (HFRG) are publishing guidance (11), little attention has been paid to how to ensure that safety-critical maintenance is properly carried out. This is relevant when the fault mode sought has rarely if ever occurred and there is little physical evidence to show that a job has been carried out. Under these conditions, checks may be omitted, partly because artisans believe that repairs have a higher priority.

Because the words 'existing' and 'good' are used together there is a risk that they are regarded as synonymous. Existing practice is not always good there are many examples of items in maintenance schedules that have at best no positive effect on safety.

2.4 Avoiding safety expenditure?

The Annual Report cites an example of QRA work that cost nearly as much as the measures suggested to mitigate the risk. It is right to highlight this risk. However, rolling stock is owned by organisations responsible for securing an adequate return on substantial amounts of capital. The techniques of risk management originated with such financial institutions and their successful creation of wealth depends on their ability to manage risk. It is hardly surprising that when decisions need to be made within a risk framework that QRA methods come naturally to the hands of the owners of the capital. Even in BR days, investment required quantitative justification. What has changed is that government cash restrictions have gone, removing the artificial distinction between capital and revenue expenditure.

It is said that QRA has been used to avoid safety expenditure. Clearly if the QRA costs a substantial proportion of the cost of mitigating the risk, its use is indefensible. The current Railway Group Safety Plan (10) also states that QRA should not be used unreasonably to justify inaction, where there are significant uncertainties in costs or benefits.

However, the reality is more complex. At the start of a QRA study one does not know the cost of all of the options, nor does one know the effectiveness of each possible measure. The aim of a risk assessment study is to identify all of the options for risk reduction and to find the most cost-effective combination of options for risk reduction. Frequently, many options for risk reduction are identified at several stages of a potential accident sequence. Each has a different degree of effectiveness and each addresses a different aspect of the risk. It is seldom the case that there is an obvious and comprehensive solution that removes the whole of the risk and has no side effects. However, the cost of QRA should be contained and be proportional to the risk and an early part of any QRA should be a preliminary quantification to indicate the cost limits for both the study and for remedial measures.

Experience shows that clients frequently commission QRA to help justify safety expenditure which they believe to be worthwhile rather than to attempt to avoid unnecessary expenditure. We are proud to be associated with Wales & West Passenger Trains Ltd who commissioned a wide-ranging study to identify and quantify potential measures for safety improvements to their fleet. While dozens of potential improvements were initially identified, application of the Pareto principle allowed the selection of a small number of potentially cost-effective measures which provided worthwhile benefits. This is in stark contrast to the previous approach; ten years ago, the industry might well have spent more on safety modifications, but would only have secured a fraction of the benefits.

There is evidence to support all of the Inspector's criticisms, but without careful reading of the report it could be inferred that QRA is out of favour and should be curtailed. In fact, the report encourages mature use of QRA and the next section illustrates the benefits by showing how techniques have been used to the benefit of the industry and its customers.

3. THE CASE FOR QRA

The arrival of QRA techniques was a boon to those who had to make decisions about system integrity. At last there was a rational basis for specifying design and maintenance standards. Commonsense has been extolled, and rightly so, but QRA contributes to creating commonsense. To provide the levels of safety required today, it is not possible to accept enough accidents for consensus to emerge. Safety management must proceed through analysis back to root causes and forward from minor accidents to potential major ones. These analyses increase knowledge more effectively than bitter experience.

3.1 Door safety

Most train door control systems used in Britain use relay logic to ensure safety. It prevents the doors from opening in motion or from closing without warning. It is difficult to create systems using relay logic which have no safety-related faults modes - that is, neither open- nor short-circuit failure will produce even minor risks. The Group Standard (12) recognizes this and requires that single-point failures should result in nothing worse than a minor injury so far as is reasonably practicable. This can be met with careful design and high-integrity components.

Things get more complicated with potentially severe faults. The Group Standard requires technical faults leading to death or major injury to occur no more frequently than every 10^{11} vehicle kilometers. This is equivalent to one major injury every 100 years on the Railtrack-controlled infrastructure. Now it is beyond current technology alone to achieve such a level of integrity, so some judgement has to be made about how humans will react to faults. This necessitates the use of event tree analysis to determine how likely it is that a combination of faults would lead to an injury.

Where practicable, circuits are designed to be self-monitoring; that is, any fault will be revealed at the next stop in a non-hazardous manner. A potential trade-off in high-integrity circuit design is between increasing complexity to protect against high consequence faults and having more low consequence faults as a result. It is no longer adequate to distinguish, as the railway industry once did, between 'right-side' and 'wrong-side' faults. The consequences of relatively frequent right-side failures, such as detrainment, may lead to a greater total risk than that posed by the wrong-side failure. Ultimately, some judgement needs to be made and a balance struck between the various possible types of fault and the associated risks. A recent project required modifications to an existing control scheme that meant there were constraints due to technology and compatibility. The calculated failure rate of a carefully designed circuit was found to be greater than before, as a consequence of the potential for failure of additional safeguards built in. Reviewers were concerned about this and a QRA was carried out to investigate the risks involved. It was found that the safety benefits of the scheme exceeded the disbenefits by three orders of magnitude, because of the different probability of injury given the various faults. Thus the design was vindicated, real benefits achieved and commonsense in door control was redefined.

3.2 To run or not to run?

Another valuable use of QRA follows safety incidents in service. Safety-related failures occurring on Railtrack's infrastructure are reported via the National Incident Room. In some cases a decision has to be made on whether to stop the fleet and carry out a campaign check. Reliability engineering methods such as probabilistic fracture mechanics (how likely is failure?) combined with consequence analysis (how bad?) can inform these decisions. They provide a means of making rational decisions, in the face of the knowledge that if the rail service is suspended, passengers will travel by means that will usually be less safe than the rail service, even taking the fault into account.

3.3 Test intervals

Historically, test intervals for protective systems such as AWS or overload relays were often selected depending on how easy the task was to do, not in accordance with how necessary it was. It was somehow seen as sensible that if a test required considerable effort then it should be carried out infrequently. In hindsight, it appears to have been thought that if inspected often enough protective systems would never fail. Apart from ignoring the physics of failure, it also neglects the human factors of maintenance; if test frequency is so high that faults are never observed, it is possible that artisans will neglect checks. One survey of AWS equipment found that 70% was unsealed despite a (scheduled) weekly check. System manufacturers often made conflicting claims for the reliability of their equipment (high) and the need for inspection during warranty (frequent).

An example of existing questionable practice is the testing of brake systems. In principle, any extension to maintenance periods requires an increase in brake test period, as brakes usually receive some attention on the most frequent examination. However, tests often focus on the highly reliable control system and do not test the less reliable actuators effectively. This was perhaps a reasonable strategy with long trains, but less so with the increased number of short, high-speed trains now in service. In consequence, while maintenance appears to control risk superficially, a deeper examination shows that the safety is ensured by the high integrity of the equipment and operator vigilance while maintenance contributes little. Yet, any proposal to increase brake test periods would surely be looked upon with disfavour. However it is possible, through disciplined use of QRA, to show that both safety and maintenance intervals can be increased by focusing the tasks on the sources of unreliability.

How then should test intervals be set? In the oil and gas industry, QRA is used to determine the required safety integrity of protective equipment. If the acceptable probability of failure per demand is selected based on consequence analysis and the reliability of the equipment is known, then an appropriate and economical test interval can be set. This is not a major, investment-based decision. It is an everyday activity for engineers working in high-integrity systems. Nonetheless, it is a rational and valid use of QRA. In practice, when reliability-centred maintenance is combined with QRA (13), such substantial reductions in both cost and risk are possible that there is little tension between the two.

3.4 Occupational health and safety

The safety of workers in the railway industry was for a long time a secondary concern. Overall, fatal accident rates were not good and those for high-risk groups intolerable. Following Clapham, the 'absolute safety' culture that swept the industry did highlight occupational safety as well as passenger safety. In some parts of BR a substantial percentage of employee time was spent carrying out safety briefings and inspecting premises.

One successor organisation was concerned about value for money of these activities and commissioned a QRA study to determine where effort should be expended on improving occupational health and safety. Some results were not entirely unexpected. For example, risks from office work were found to be well controlled and business travel by car was found to be a greater risk than working on the trackside for this organisation. However of rather more significance was the fact that stress-related illness was a thousand times more important than any other risk. This finding was only possible because of quantification.

A similar exercise using risk matrices had failed to highlight the state of affairs. It is easy to forget that on most matrices, there are ten or more orders of magnitude difference between the two corners (Figure 1). The figures can be regarded as either risks or the amount of money that should be spent on mitigation.

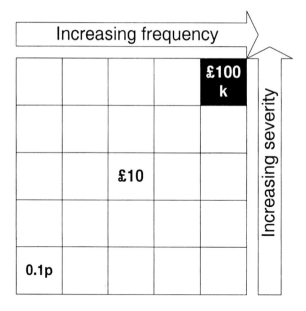

Figure 1 A typical risk matrix with risks quantified

Without quantification, engineers are seldom ruthless enough in discounting the minor risks, nor do they always appreciate the gravity of major ones. The client organisation can now continue to invest in employee care, but can focus on the significant issues.

3.5 Market forces

It is true that misuse of QRA can produce any answer that is desired. And, while this criticism can be levelled at any analytical technique, QRA is potentially more subject to abuse because the events being analysed are infrequent. The collection and application of data and the construction of models can seldom be tested against reality for this reason.

Professional engineers are subject to market forces and those who continue to apply the old BR 'cost no object' standards to their professional work will lose business. Clients will not tolerate engineers who fail to deliver on time or to budget because of vague concerns about imponderable risks. However, professional engineers do not idly dismiss or cover up risks because of a client's desire to save money. Nor, it is contended, do clients commission work with anything other than a proper regard for gaining value for money from safety expenditure.

Many engineers accept without a second thought the 1/100 risk that they (as a motorist) will kill somebody one day. They appear to have disproportionate scruples about carrying out work within the discipline for which they may have decades of education, training and experience. It is a sad reflection on the engineering profession that a relatively safe and environmentally sound mode of transport is made unaffordable to many by engineers who do not have the deep understanding and effective tools necessary to make good professional judgements on risk.

Many voices in society highlight the qualitative and sociological aspects of the acceptable risk debate. It is necessary for engineers and the railway industry to participate in and inform this discourse. It is necessary to balance those legitimate concerns with a reminder that society's prosperity, its ability to afford improved material standards, is contingent on the sensible economic management of risk.

4. CONCLUSIONS

QRA has given the engineering profession the tools to set costs and benefits in perspective, to improve rail's competitive advantage while improving safety. It would be unfortunate if that enterprise were frustrated due to misunderstandings over the value of diligent QRA.

QRA does not inevitably lead to a decline in safety standards. The opposite is the case - it is common for careful analysis to show that risks are greater than perceived and that extra measures are needed. It is accepted that QRA has on occasion been used in ways that are far from best practice. However, it is believed that the exposure of assumptions to scrutiny made possible by the transparency of good QRA is of immense benefit in improving railway safety and demonstrating value for money.

Society demands that railway safety be further improved. Costs have to be reduced. These are the imperatives of the marketplace, regardless of government shade, for the twenty-first century. QRA, used to inform societal and engineering judgement, offers a major contribution to their achievement.

REFERENCES

1 *Railway Safety: HM Chief Inspector of Railways' Annual Report on the safety record of railways in Great Britain during 1996/97* London: HSE Books 1997.

2 *Transport Statistics, Great Britain*. DTp, HMSO, London 1996.

3 *Ensuring Safety on Britain's Railways*. DTp, London, Jan 1993.

4 *Railway Safety Cases: A summary of contents and risk assessment*. AW Evans and AX Horbury. Centre for Transport Studies, University College, London, 1997.

5 *Railway Group safety data companion version 1.1*. Railtrack S&SD, March 1997.

6 *Railtrack's Railway Safety Case volume 3J: Major risks within Railtrack Zones*. Railtrack S&SD, London, 1996 (since superceded by volume 2, section F of Issue 19).

7 *The tolerability of risk from nuclear power stations*. HMSO, London, 1992.

8 *Proposed railway safety regulations 1998: Cost benefit assessment*. HSE Economic Advisers Unit, London, May 1998.

9 *Response to the Heath and Safety Commission consultation document 'Rail safety: proposals for regulations on train protection systems and Mark 1 rolling stock'*. AW Evans University College, London, August 1998.

10 *Railway Group Safety Plan 1998/99*. Railtrack, London 1998.

11 *Management of Maintenance Error*, pre-publication draft 7, 1998. To be published by HSE.

12 Group Standard GO/OTS300 Issue 1. *Power operated external doors on passenger carrying rail vehicles*. British Railways Board, London, 1993.

13 'Service Driven Maintenance', JT McMenamin in *Railtech 1998 Train Maintenance*. Professional Engineering Publications, London, 1998.

C552/037/98

Group standards and vehicle acceptance

A SUTTON BSc, MSc, MRAeS
ABB Daimler-Benz Transportation, Derby, UK

The experiences of a vehicle builder moving into, and working within, the Railway Group Standard environment are reviewed. The significant issues are addressed from the point of view of the expectation of the customer, the vehicle designer and manufacturer and a Conformance and Vehicle Acceptance body. It gives a view of the key changes that have occurred to the standards and associated procedures and the practical implications. The paper concludes by summarising progress made to achieve the current position and some of the outstanding issues.

1. INTRODUCTION

This paper discusses some of the experiences of Adtranz Rolling Stock, as a vehicle designer and manufacturer and as a Certification Conformance Body and Vehicle Acceptance Body, when working in an environment incorporating Railway Group Standards. It addresses the requirements placed upon the company, the issues that have arisen and how they have been addressed. Some problem areas and solutions are discussed. It also looks to the future and examines changes that are currently being considered.

2. BACKGROUND

Before discussing the Railway Group Standards (RGSs), it is necessary to understand the environment in which they exist. They are a practical expression of the process of achieving a safe railway system through the Railtrack Safety Management System. Figure 1 illustrates where vehicle Engineering Acceptance, with which we are now concerned, fits into the overall safety management process. The need to acquire Engineering Acceptance via a certification process is a parallel requirement to achieving Route Acceptance through an operational safety case. Acceptance by Railtrack is a prerequisite for HMRI to give statutory approval for the operation of vehicles.

From the perspective of a vehicle manufacturer, Railway Group Standards should define all the relevant interface requirements between the vehicle and its operating environment such that it can be operated safely on Railtrack's infrastructure. These requirements may be physical (e.g. track gauge), performance related (e.g. braking distance) or operational (e.g. driving sight lines and controls). Also included within the scope of RGSs are procedural standards that define the acceptance process itself.

In general, Railway Group Standards do achieve the above requirement. However, it should be recognised that the scope of vehicle standards has been reduced from a set covering many aspects safety, to ones associated only with the potential impact on the infrastructure or other rail users. Other requirements, that are not interface related, but for which there is a need to have consistency from both a safety and an industry efficiency point of view, such as the structural integrity of vehicle interior fittings, are no longer included. An alternative method of managing safety standards in these areas has still to be established.

The overriding objective behind the Railway Group Standards must be to make a positive contribution towards assuring a safe railway at an affordable cost. If either of these objectives is lost then our industry's future will be under serious threat.

3. MANUFACTURER'S EXPERIENCE

3.1 Conformance

The products we manufacture must conform to the Railway Group Standards. Since the RGSs have as their origins the BR specifications, in principle this should be little different to meeting the essential requirements formerly specified in BR contract documents. In practice, there are significant changes.

One of the first issues to emerge was that the RGSs often specified more rigorous requirements than had previously been achieved. These requirements may have been in the contract specifications in the past, but had never been fully achieved, and had either been modified at the contract stage or resulted in a concession when full compliance was demonstrated to be impractical. A typical example would be the current interpretation of the requirements on the size and shape of an obstacle deflector, which is expected to be much larger than previous designs and to extend well beyond the normal vehicle profile.

A major difference in the application of RGSs is that the opportunity to be 'non-compliant' to a requirement and negotiate a mutually agreeable alternative solution no longer exists in practical terms. The process for addressing a potential non-conformity now requires a formal approach to Railtrack, which is at arms length from the problem, and separated in the normal course of events by its appointed certification agents (CCBs and VABs).

It is quite appropriate that RGSs should be non-negotiable. However, the difficulties with some of the early standards, in terms of both requirement and interpretation, necessitated discussion, and the route for any such approach to Railtrack was initially poorly defined. It was not so much a case of not being able to address such issues, as that the time required to resolve them had serious repercussions on contractual obligations and hence on project costs.

Another point very relevant to the manufacturer is the date of application of a standard and the policy when the standards change during a contract. For some time there has been a common sense understanding that the standards in force on the day a contract is signed apply for the duration of the contract (unless the parties agree to adopt a subsequent issue). Figure 2 illustrates a possible sequence of events where this is relevant. However, a clear statement that recognises this, and which permits modifications necessary to meet the terms of the contract to be compliant with the original applicable standards, and not subsequent revisions, has still to appear in the documentation. This problem has now been recognised by Railtrack and a joint drafting group is being set up with industry representation to produce a clear application document.

3.2 Standards Development

BR/Railtrack engineers produced the original series of RGSs in a very short period of time. This was no mean feat and should be applauded, especially when it is realised that a vast number of BR standards and specifications were reduced to the current more manageable number. However the timing and speed of this process prevented outside participation or effective consultation and the results were inevitably a less than perfect set of requirement documents.

Railtrack has a programme of progressively reviewing and improving standards and has actively sought the involvement of the supply industry in this process. Adtranz believes that full industry participation is essential if the standards are to reflect the full knowledge base and the practical and commercial implications of changes are to be fully understood and the benefit accurately assessed under the ALARP principle. It actively supports the standards development process by participating in drafting groups and commenting constructively on all the relevant draft documents it receives.

There are still key areas where the information contained in RGSs is inadequate. Two of the most important interfaces lacking clear definition are in the maximum permissible vehicle swept envelope (swept gauge) and the signalling system electrical interference tolerance limits. In considering these issues it is important to remember that the market is not interested in buying vehicles with any performance or route restrictions. In the privatised railway the value of a vehicle is closely related to its capacity and its long-term flexibility of use.

At present, gauge clearance assessment relies primarily on comparison with existing rolling stock. Successive gauging by such a method must logically result in progressively smaller and smaller vehicles, as the accuracy of data on the reference vehicle is usually uncertain (what were its build tolerances, for example) and a safety margin should be included in each iteration. This is clearly unacceptable as a long-term strategy. There are many questions which need answering in this area, starting with fundamental points such as who (vehicle manufacturer or infrastructure builder) should take account of what at the interface. An extension to the general uncertainty emerged when, in searching for data to define the actual available clearance at platforms, it was surprising to find that Railtrack track positional tolerance limits appear only in a Code of Practice and not as a mandatory requirement in a RGS.

The problems that have been encountered with the potential interference of signal circuits caused by modern ac traction equipment are outside the scope of this paper. However, they represent a safety critical vehicle interface issue and the requirements, both in terms of vehicle performance and the method of verification should be clearly defined within the RGS regime.

As a result of the efforts of Railtrack and the industry, the general situation regarding RGSs has now improved on all counts. The involvement of industry in the drafting and re-drafting of standards has been a major contributor to this improvement. This participation has also been beneficial in helping to bring the industry together in a common objective, following the disruption caused by the restructuring. It is unfortunate that the participation of outside parties in standards activity is still only at Railtrack's discretion. There is no mandate, as with national and international standards authorities, to invite representation from all interest parties.

3.3 Internal Issues and Processes

Not all the issues resulting from moving into a RGS environment are associated with Railtrack or the standards themselves. There has had to be a significant culture change within our company. It has always been Adtranz policy to design safe trains; not to make the trains it designs safe. However, the formal audit trail for demonstrating safety through conformance to standards has changed from a largely customer driven process to one which the contract project team must manage itself. To accommodate this change Adtranz has taken a number of actions.

It is important to note that showing conformance to the engineering RGSs requires nothing in addition to that which should be done anyway to ensure that the product is safe, functionally fit for its purpose and an acceptable commercial risk for its manufacturer. The new demands imposed by RGSs are associated with changes in process, timing and presentation. Adtranz has found it advantageous to revise and develop its internal procedures and processes to conform more closely to the demonstration requirements demanded by the vehicle Engineering Acceptance process. These changes have been augmented by staff awareness sessions to ensure that those involved understand the Engineering Acceptance process, its expectations and how the correct information can be best presented at an appropriate time and in the most effective format.

The challenges we have had to deal with in-house have also been presented to our equipment suppliers. It cannot be emphasised enough that providing the information for scrutiny is just as important a deliverable as the equipment itself. A Scrutiny Certificate of Conformance cannot be issued until **all** relevant items have been covered, and without one a vehicle cannot run, even on test. By its nature, completion of scrutiny will always take a lot more time than fitting the corresponding part, and the delivery schedule must recognise this!

3.4 Interface with Route Acceptance

Timing of the certification process has been, and still is, an area we have to satisfactorily resolve in another respect. The Railtrack RSAB process assumes a staged progression of a project through concept, design, manufacture, test, and introduction into service. There is a sense of following a consecutive series of events. This does not reflect reality in the present UK railway contractual environment. Time scales dictate that most of the above activities have a large degree of overlap or happen concurrently.

The Design-stage submission within the Railtrack RSAB procedure reflects such a timing inconsistency. In theory this submission should be accompanied by a Design Scrutiny Certificate (see figure 3). The requirement does not recognise that a Scrutiny Certificate cannot be issued until **all** relevant RGS requirements have been answered satisfactorily and that this will not usually happen until the first vehicles of a build are complete and several others are in advanced stages of production. The design and test conformance milestones occur virtually together.

One can see the logic in using the Design Scrutiny Conformance certificate as evidence that hazards addressed by conforming to the RGS requirements have been eliminated. However, since a product cannot be registered on the Rolling Stock Library until it has a certificate of Engineering Acceptance, and hence must conform to all relevant RGSs, the need to demonstrate the inevitable in advance seems an unnecessary as well as impractical requirement. Means of addressing this conflict are currently being explored.

4. CERTIFICATION CONFORMANCE AND VEHICLE ACCEPTANCE BODY'S EXPERIENCE

4.1 In-house Certification

Adtranz decided to apply to become both a Certification Conformance Body and Vehicle Acceptance Body as soon as it became clear that it was eligible to do so. The reason is obvious: it wished to be able to manage the acceptance process for the vehicles it produces in the most effective manner. Timing is again of the essence. There is also no better way to understand a process than to become actively involved in all its aspects.

When Adtranz was accepted as a CCB and VAB the representative of Railtrack who was present said, 'Welcome to Railtrack'. As these bodies are effectively Railtrack's (and hence the standards authority's) agents, this remark was very significant and must be remembered by all who are involved in their activities. It is, however, one of the very few references that have recognised the status of these bodies. The revised Railway Group Standard Code does not acknowledge their existence, let alone identify their status.

Status may seem relatively unimportant, until it is realised that it seems to determine whether you are able to receive **controlled** copies of RGSs (which is **very** important) and what you have to pay for them (a topic close to every manager's heart).

Adtranz can achieve and demonstrate the necessary degree of independence required by a Certification Conformance Body due to its internal 'one roof' project management policy. Project teams (including the design function) responsible for meeting specific contracts, operate within the company as effectively independent entities. Engineers not part of a particular team can, therefore, assess its output with the independent objectiveness necessary to perform the certification function. The Adtranz experience to date is that internal scrutineers tend to be more thorough than third parties: perhaps because they have a better insight into the processes involved and a greater awareness of the need for impartiality. Other organisations have reported similar experiences in this respect.

Because of the scope of its operations and the corresponding expertise within it, Adtranz was able to obtain approval as a Certification Conformance Body in all five current fields, plus that of Engineering Acceptance (VAB). In addition to its designing and manufacturing expertise, it already had a development test laboratory familiar with the testing of vehicles for BR and capable of performing all the static and track tests required by the RGSs. Adtranz Customer Support Division has the necessary expertise to certificate maintenance and overhaul policy.

The Certification Conformance Body must remain independent of the project team. However, if the objectives of both are to be met they must communicate. Achieving conformance must be an integral process within the project from its start and a matching input from the CCB is required. With a new vehicle, the Design Scrutiny task requires in excess of 3000 individual requirement items (clauses) to be confirmed. With a task of this size the initial planning is as important as the final execution to both project and Conformance Body. The associated resource management task is equally critical, typically requiring the planning of more than thirty man weeks of work incorporating a range of skills and experience. For a large build of vehicles, production may continue for several years. The Manufacturing, Safety and final Acceptance certification of each vehicle/unit as it is completed must likewise continue and be managed over this period.

4.2 Practical Issues

An area of some uncertainty is the scope of responsibility assigned to the CCBs and VABs. We are never likely to be in the ideal state of having fully comprehensive and unambiguous standards. What degree of interpretation can a CCB apply, when should it refer to the VAB and when is Railtrack S&SD to be consulted? Some of the uncertainty may be due to a careless use of terms. For example, it is often quoted that the VAB is responsible for assessing the acceptability of ALARP justifications. But in the normal course of events it is the CCB that examines the design in detail and is in a better position to make such an assessment. The VAB's prime function is to audit the processes applied by the CCBs and ensure that it is carrying out its responsibilities correctly and with due diligence. It would be helpful to have better defined scopes of responsibility.

The in-house CCB/VAB has proved to be of particular benefit when a new vehicle type is going through final assembly and acceptance testing. As manufacturing processes are tuned and performance data is accumulated, a progressive updating of details in the vehicle design, its assembly and validation programme takes place. This in turn can have a sequential knock-on effect on the permitted performance envelope during the testing phase. Such a sequence of events can require amendments to the vehicle certificates in the form of endorsements or re-issue. To minimise delays to the project work programme, a continuous liaison and rapid response by the CCB/VAB is essential.

There are still a number of areas where procedures for the issue of certificates and registration on the Rolling Stock Library need clarifying or improving. The logistics of issuing and processing certificates when vehicles are being produced at a high rates (e.g. one a day) and need to leave the site at night or over a weekend is one such example. The present procedures must be developed to provide the necessary level of responsiveness and to address the need for 'out of office hours' clearance.

5. THE FUTURE

5.1 New Group Standard Code

When, earlier this year Railtrack published its proposals for a new Railway Group Standard Code it was a disappointing document for the manufacturing industry in many areas. However, the subsequent response of Railtrack to the industry comments has been very encouraging and a much better final document is expected to appear. In particular the proposed level of representation on the Subject Committees from outside the Railway Group now more closely reflects the emerging balance of knowledge within the industry. However, as noted previously, the commitment to involve organisations outside the Railway Group in the drafting of standards is still expressed as being wholly at the discretion of Railtrack.

One very positive proposal in the new Code is to set up a top-level steering committee to oversee the development of RGSs. It will provide balanced and objective guidance, with specific reference to maintaining and improving safety, by drawing its members from all fields of the industry. As required by the present legislation, Railtrack remain the safety management authority. While in no way questioning the integrity of the Safety and Standards Directorate, the manufacturing industry consider that a better long term objectivity would be achieved if the standards management authority was an independent body (as in other fields of transport).

5.2 Less Prescription

Railtrack has declared an intent to move towards less prescriptive standards. Requirements will be expressed in the form of performance/risk objectives. This may be a mixed blessing for the vehicle manufacturer. Prescriptive standards are essential to control interfaces. In their formation they also incorporate a high degree of collective knowledge and experience. A performance/risk analysis is only as good as the expertise of the assessor and hence this approach has the potential for increasing risk and reducing safety.

From a practical point of view, the potential advantage of having greater design flexibility, as offered by a less prescriptive standard, may be more than offset in most cases by the extra work that must be done to demonstrate a less precisely expressed conformance requirement. The supply chain also needs a relatively high level of prescription if it is to respond effectively. The favoured route would be to retain the proven solution as a prescriptive requirement, and permit the option of an alternative approach supported by a comparative performance or risk assessment.

5.3 Revision of Certificates

The latest proposals for changes to the Engineering Acceptance process reduce the number of certificates required, before the VAB issues the final certificate, from five to three. The Type Test certification is to be integrated into Design Scrutiny and the Safety Examination is no longer to be a separately certificated event.

It is certainly appropriate that the Safety Examination should not be a separate, distinct part of the certification process since it is a necessary conclusion to any major work on a vehicle. The first safety check may be considered as the end of the manufacturing process and subsequent ones as a conclusion to a repair or maintenance schedule. As such it should be part of these operations.

A similar justification can be made for the proposed change with respect to Type Testing. The present Design Scrutiny process is intended to provide confirmation that the design, on paper, conforms to the required standards. The Type Tests provide the physical proof of conformance in key areas. In practice this distinction is not always clear, as some areas are common to the Scrutiny process and, where the calculations leave uncertainty or show a marginal conformance, the Scrutiny CCB often stipulates that conformance should be validated by testing. Testing can logically, therefore, be considered an extension of the Scrutiny process and, as stated earlier, has little separation from it in time. As far as the issuing of certificates is concerned, it must be recognised that a number of successive Design Scrutiny certificates or associated endorsements are issued as the vehicles progress through these stages of the acceptance process. The current RSAB staged acceptance procedure, as indicated in figure 3, will also need to comply with this reality.

5.4 Cost Implications

As with other aspects of rail privatisation, the current vehicle acceptance process has changed the allocation of the associated cost and risk. These were formerly largely hidden within the BR organisation, but have now become a clear contractual item. The direct cost arising from the assessment of design and manufacturing records and the issue of certificates is significant in itself. Any risk or uncertainty associated with the process or its requirements of necessity incurs additional indirect costs. If the requirements are excessive or ambiguous then the contractual risk and cost is increased. It is very important, therefore, that standards are appropriate, clear and practical so as to ensure that acceptable levels of safety are achieved whilst at the same time minimising the cost to the industry as a whole. The proposed reduction from five to three areas of certification is a positive step in this respect.

The management of safety in areas no longer covered by Railway Group Standards urgently needs to be addressed. The industry needs to have standards to work to if the best collective knowledge is to be applied to safety in these areas. Failure to resolve this situation will result in the continual expenditure of additional time and cost which the industry can ill afford.

6. CONCLUSIONS

The Railtrack Engineering Acceptance process is a key element in its method of achieving the safe operation of trains on its infrastructure. The Railway Group Standards form the core of this process and must provide the necessary definition of the interfaces between the vehicles and the infrastructure in order to enable safe trains to be designed. After some initial problems, a workable system of standards has emerged. Adtranz finds that this new set of standards provides a generally agreeable environment in which to design vehicles.

Moving to RGSs from the previous working practices with BR presented a number of problems mainly associated with loss of direct contact between the parties defining and responding to the requirements and the movement of knowledge within the now privatised industry. The early difficult situation improved as the industry became used to the new environment and a wider participation of all interested parties was achieved in the development of standards. This participation needs to be extended if the consequences of changes to standards are to be fully understood and taken into account, particularly at the interfaces between traditionally separate disciplines (e.g. civil and vehicle engineers).

Railway Group Standards also provide the framework for demonstrating effective control of vehicle manufacturing and maintenance functions. The main demand on the manufacturer in this area has been to adapt the in-house disciplines and audit trails to be consistent with the specific RGS requirements rather than to introduce any fundamental changes to processes or functional activities.

A number of key interfaces are still inadequately defined. These potentially add risk and cost to the design and operation of vehicles and/or impose otherwise unnecessary constraints. Adtranz wishes to see these issues resolved as quickly as possible and will work with the rest of the industry to achieve this objective.

Even though it carries much of the responsibility for implementing them, under the present arrangements the supply side of the industry is permitted a less direct influence on standards development than the operating companies (Railway Group). This state of affairs is at variance with normal procedures for standards development and must be detrimental to the overall effectiveness of the process.

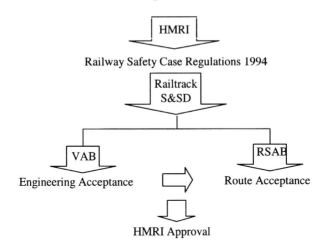

Figure 1. Safety Requirement

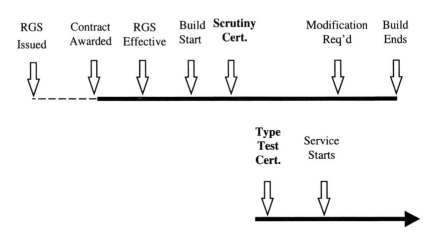

Figure 2. Possible Sequence of Events

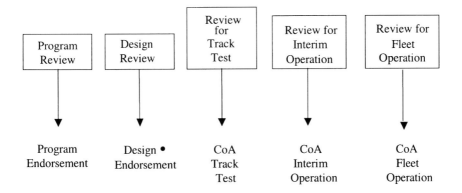

Figure 3. RSAB Acceptance Stages

C552/011/98

Commissioning trains

R HARDI
ABB Daimler-Benz Transportation UK Limited, Derby, UK

1. SYNOPSIS

The modern and especially the privatised railway industry is demanding from manufacturers ever shorter lead times, lower prices and better performance whilst expecting one hundred percent availability on the first day of introduction into service.

These new demands have a major impact on the industry and have forced manufacturers and service organisations to re-think their strategies. This paper presents the case for the inclusion of commissioning along with other modern manufacturing concepts such as "out of the box" concepts and "simultaneous engineering" approaches.

The commissioning stage, the "Link Between Build and Service" is the first and in many instances also the last opportunity for the manufacturer to prove and test their designs under service conditions and gain valuable feedback for current and future products.

The requirements and experience of commissioning should be included in a project as early as in the tendering phase and been carried through even beyond the service introduction phase. Apart from the normal testing to ensure that the product is manufactured to the expected quality, it is vital that performance and reliability data is fed back to the engineering process during the commissioning process.

2. THE DEFINITION OF COMMISSIONING

As with manufacturing the role of commissioning must also evolve to meet the new requirements, hence the process needs to be properly defined.

2.1. Definition

In the "Oxford Reference Dictionary" commissioning is described as "to prepare for active service". In the context of rail vehicles, commissioning therefore is the process of preparing rail vehicles for passenger, freight or special purpose service. It implies that there is a defined period between the production of a unit and its introduction into service. Due to the intensity of activity during the build up to active service it is important that commissioning is integrated into the entire production cycle.

3. INTEGRATION IN THE PROCESS

Depending on a company's structure, commissioning can start at several stages in the production process. The deciding factor must be whether the process can benefit from the experience of the commissioning department.

3.1. Tender stage:

The inclusion of commissioning knowledge in the tendering phase utilises a valuable source of service experience. It can help in the process of minimising risk and optimises cost calculations. The number of test requirements from train companies has increased significantly and the implications to cost and production schedule have to be considered in the tendering phase.

3.2. Design & build phase:

The same principles apply to the design and build phase. The inclusion of commissioning personnel at this stage allows the feedback of valuable knowledge from previous builds to be integrated into the design. It is also the ideal opportunity for the commissioning staff to familiarise themselves with the product. The integration of selected commissioning engineers in design reviews can provide fresh inputs from a different angle. (this follows the concept of concurrent engineering)

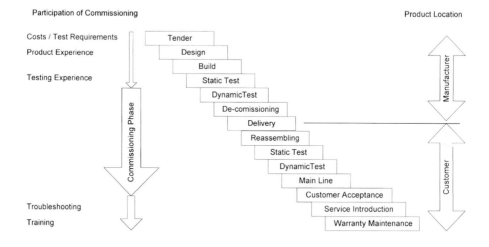

Figure 1: Main stages of the production process

3.3. Static testing:

At this phase the vehicle design should be completed & all subsystems functional, only at this stage can static testing be started. The commissioning specification and the static test specification have to be harmonised to eliminate duplication of work. This is the most appropriate stage for handing over the vehicle to commissioning.

3.4. Commissioning stage:

This is the phase where all fitness for service testing is conducted prior to entry into service and hand over to the customer. Commissioning starts with first movement tests and ends with main line running tests. It is important that this phase is not compromised and used as a buffer to make good lost time during production.

3.5. Service introduction & warranty stage:

The people involved in commissioning are ideal to provide the needed security on service introduction and for transferring the new product knowledge to the train maintainers. Whilst the new owners familiarise themselves with the new product, the integration of already trained commissioning engineers can ease the sometimes hectic introduction period. Whilst this is happening – in many cases by default – an early recognition of such need promotes better planning and therefore more efficient knowledge transfer scenarios. This process can also be shortened, by integrating maintenance personnel into the commissioning team.

3.6. Type testing:

During the type-test phase the commissioning department has a clear role to keep the vehicle functional allowing valuable engineering resources to focus on the issues at hand.

4. INTERFACES

Dependent on the company's project organisation, commissioning could start anywhere in-between static testing or dynamic testing. At whichever stage the hand-over takes place, suitable procedures and documents will guarantee that all requirements are met.

4.1. Production

Detailed hand-over documentation incorporating signed off testing specifications and configuration status help to define the condition and functionality of the vehicle at hand-over hence ensuring its inclusion in any future configuration changes.

4.2. Engineering

This two-way interface allows knowledge of vehicle design to pass to the commissioning department allowing the commissioning to take place and field experience from commissioning to be fed directly back to the design departments.

4.3. Maintainer & operator

After successful completion of commissioning, the vehicle is handed over to the maintainer and/ or the operator. This hand-over has to be an agreed process, usually forming part of the acceptance of the vehicle by the customer. In many cases the signed commissioning specification is sufficient to act as the hand-over document.

5. THE COMMISSIONING ORGANISATION

The previous chapters have described the various processes of, and interfaces to and from the commissioning department, it is important that the structure of a commissioning organisation reflects these requirements.

5.1. Organisation

The organisation centres on the concept of a core of dedicated commissioning engineers that can be moved from project to project and integrated into a project taking their knowledge and experience with them. While they are integrated in a project they remain connected to a central commissioning department. This set-up enables the project managers to maintain their direct control over project issues while continuity of commissioning quality is maintained through the commissioning department.

5.2. Commissioning engineer

The diverse tasks that comprise the commissioning of a railway vehicle demand a highly disciplined multi-skilled person. The commissioning engineer must understand all vehicle systems and have the skills to feedback information to the engineering function, sub-contractor and customer in a appropriate analytical systematic manner. Due to the nature of commissioning, commissioning engineers have to work extended hours, weekends and at many times under arduous conditions. People who combine the needed skills, and prepared to work under such conditions are difficult to find and their employment conditions must reflect the special requirements if suitable engineers should be attracted.

5.3. Management

By necessity because of the integration of the commissioning teams into projects, management must become a joint task between the commissioning manager and the project manager. The importance of genuine integration of the commissioning teams within the individual projects can not be over stressed.

6. DOCUMENTATION

Well-structured documentation allows continuity of process across projects and guarantees standardisation throughout the company. There are a lot of different documents, which are important for commissioning, of which probably the most important ones are the commissioning plan and specification.

6.1. Commissioning plan

The commissioning plan defines the different stages in commissioning and must be integrated into the overall project plan. The individual stages should also be discussed and agreed with the operators since they may interfere with daily routine.

6.2. Commissioning specification

The specification defines what functional testing is required at each stage of the commissioning process. It should be designed in a way, that it is easy to understand and can be used as a reference document.

It should define the required actions, processes, amount and necessary conditions of testing for each function of the vehicle.

Vehicle functionality is broken down into discrete parts, each of which is tested and signed off as part of this process. The responsible commissioning engineer in conjunction with the relevant design engineers designs the commissioning specification. Every functional test should be quantified i.e. each test should have a clearly defined result (pass fail criteria).

6.3. Commissioning specification example

As example the Adtranz 168 Chiltern commissioning specification describing the door test:

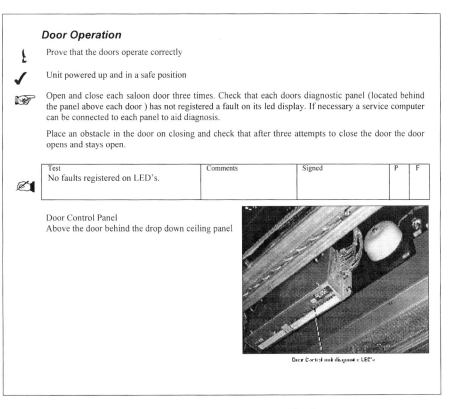

Figure 2: Extract of a commissioning specification

7. FEEDBACK REPORTING

A feedback system plays an important role in the design process – acting as a reporting mechanism system for design issues over the entire product production cycle. During the commissioning phase the vehicle will undergo its first movement tests and hence it is vital that the feedback system is already in place and well understood during this crucial period. The data sent via the feedback system can take many diverse forms: documentary, pictorial or electronic. Electronic feedback encompasses all on board vehicle diagnostic systems recording vehicle metrics where applicable this data can be fed back into the design process.

8. SUMMARY

This paper presents a case for a new look at the sometimes-overlooked function of the commissioning process. The paper supports the argument that commissioning is not an isolated process and presents the need for integrating commissioning into the concurrent engineering of the entire production process.

The people responsible for commissioning should been given the tools and knowledge to provide the required service.

C552/096/98

Reliability growth in the rail industry

M H IRWING MA, MSARs
Adtranz, Derby, UK

This paper examines the process of in-service reliability growth of trains. With reference to case studies it shows how the success of reliability growth is heavily influenced by the realism of the target, the basic design of the vehicle and choice of subcontractors and the availability of good reliability data.

1. INTRODUCTION

Reliability growth is an engineering process whereby the reliability of something is improved with time. There is a school of thought that says that reliability growth should be an "off-line" activity and that products which are actually delivered to customers should be reliable from day one. In many industries, particularly where there are high production volumes, this is a realistic expectation. However, whilst Adtranz are making every endeavour to make this happen in the future, the present situation in the rail industry in Great Britain is that much reliability growth occurs whilst the trains are actually in-service. It is this period of in-service reliability growth that is the main subject of this paper.

The main thesis of this paper is that successful reliability growth depends upon three main factors:-

1) A realistic reliability target which all parties understand and are committed to;

2) A design of vehicle and choice of subcontractors which forms a sound basis for achieving reliability;

3) Availability of good reliability data so that problems can be identified and solved.

This paper will explore this thesis with reference to three case studies; the Class 465, Class 325 and Class 168.

2. CASE STUDY 1 The Class 465 "Networker" EMU

2.1 The Context

The Class 465 EMU was one of the last BR procurements. It was an aluminium bodied EMU with AC traction, sliding plug doors and a number of other features that were innovative at the time.

2.2 The Target

The target reliability was set at 60,000 miles between technical casualties, which was better than much simpler vehicles such as Class 319's were achieving at this time.

At the time of this contract I was working for British Rail. Taking account of the complexity of the units, I calculated at that time that in order to meet 30,000 miles per technical casualty, BREL (as it then was) would have needed to have selected the most reliable currently available supplier for each subsystem. In other words, achieving the reliability target would involve a substantial improvement over the current standards.

The reliability contract was based around achievement of the 60,000 miles with an assessment period during which failure to achieve the target resulted in penalties. In accordance with the BR procurement method at the time BREL were heavily pressurised to accept this contract. However, because at this time vehicle builders did not have routine access to reliability data for their fleets once they were out of warranty, it was very difficult for them to assess how realistic the figure was or what would be entailed in meeting it. Even when BREL did get some reliability data it was still difficult to assess because the quality of the internally generated BR data was not particularly good and BREL would probably have ignored the many cases where no hard fault was identified thus giving an optimistic view of the current reliability at that time.

2.3 The Starting Point

The sub-contractor for the traction system (Brush) was specified by BR, although this system had only run for a short period as a prototype. This was proven to be a problematic choice and BRs involvement in this made the relationship between BREL and Brush rather difficult.

The traction system had to be shown to be incapable of generating interference with signalling and in order to do this an Interference Current Monitor Unit (ICMU) was fitted. Unfortunately, the interference levels specified by BR were so low that the ICMU was continually tripped by arcing shoegear and other trains generating interference.

The door mechanism, pneumatic control and electric control were from three different organisations. Thus, when door problems arose it was always difficult to identify the supplier responsible.

Several other systems were of a type which had caused problems in the past but it was hoped that the subcontractors would rectify these known problems.

In design, efforts were made to ensure that previous design errors were not repeated by having an After Sales Engineer on the project team with a remit to feed in after sales experience. Unfortunately, this effort was only partially successful.

When the 465's entered service in 1993 the reliability was initially poor. Major problems were experienced with traction, doors, wash wipe, couplers breaking and toilet failures leading to water ingress into body end jumpers. Two systems (the traction system and auxiliary inverter) were badly affected by their own inbuilt diagnostic systems which were too sensitive and caused the systems to shut down spuriously. Neither of these systems was easily reset following such shutdowns.

None of the deficiencies was so fundamental that it could not be solved with realistic expenditure of money and effort and hence the vehicles must be regarded as a reasonable basis for the achievement of reliability growth.

2.4 Management of Reliability Growth

As stated before the Traction System and ICMU were major causes of unreliability. However, because of heavy BR involvement in these systems it was difficult for BREL to effectively manage the situation.

In relation to toilets, ABB (who had taken over BREL) was convinced that the problems were due to customer abuse (e.g. putting coke cans down the toilet). BR argued that this was expected use. At time of contract ABB had suggested a more expensive toilet would be better but BR had chosen the cheaper system and therefore were implicated in the problem. Eventually the toilets were replaced with those from another manufacturer.

Further problems with train availability were caused because BR's requirement of having no visible fastenings meant the interior was not very vandal resistant and, when damaged, panels were hard to change. To some extent this involvement of BR in the problem caused management to focus on debating the reasonableness of the BR requirement and what constituted vandalism rather than on solving the problem. What was perhaps particularly galling was that the GEC fleet of Networkers delivered at the same time did use visible fastenings even though BR had been very firm that BREL should not do so!

A major issue which mitigated against reliability growth was the level of information in reports from the operator following delays. In many cases the sole information would be something like 'delay 5 minutes. Unit 3 – Door Problems'. There would be no description of why this caused a delay or how the train was made to move again. When ABB were given the train to investigate they would rarely find a hard fault. An example of the reports is presented in Table 1.

A final issue regarding reliability growth management was that because the maintenance was being carried out by the operator rather than ABB there was always some doubt about whether it was being done correctly, particularly in some areas like doors. This meant that there was a high percentage of failures which were attributed by ABB to poor maintenance. As an experiment, ABB re-set a number of doors and marked all the adjustments with paint. When the unit came back for its next scheduled maintenance every door had been 'readjusted'. Against this background it was very difficult for ABB to understand whether there were any design problems with the doors or not.

2.5 The Result

The reliability growth curve for the Class 465 is presented in Figure 1. From this it can be seen that the Class 465 was achieving 4000 miles per casualty in August 1993. This increased to 30000 miles per casualty by April 1994. However, the reliability then dipped in the summer of 1994, a seasonal trend which has continued to this day. The target of 60000 miles per casualty was finally achieved in March 1996.

Unfortunately there has been a difference of opinion between the train operator and ABB about the "real " reliability of the trains. The train operator sees every delay to the train, whatever the reason, and this impacts on the ability to run a service. The train builder, particularly bearing in mind the adversarial nature of the BR contract, only feels responsible for those failures which can be reasonably shown to be directly his fault. Bearing in mind the difficulty in obtaining full and accurate data on delays in the running railway and the large number of possible causes of delays this difference can be difficult to resolve. However in more recent contracts between the two parties this has been recognised and an attempt has been made to achieve a greater degree of harmony.

2.6 Conclusion

It is hard to assess whether the reliability target was realistic since, as has subsequently become apparent, the data that was being used to monitor reliability in BR at this time was somewhat suspect. However, what is clear is that the lack of easily available reliability data in the industry meant that the understanding of both the main contractor and the subcontractors must be doubted.

The basic vehicle design and choice of subcontractors formed a reasonable basis for reliability growth.

The management of reliability growth was hampered by BR involvement in some of the key decisions which made it difficult for the main contractor to manage reliability.

The natural difference in outlook between the operator and the train builder coupled with the difficulty in obtaining full and accurate reports of service delays has created some disagreement about the reliability of the trains. This has been recognised by both parties and future contracts between them do address this issue.

3. CASE STUDY 2 The Royal Mail (Class 325) EMU

3.1 The Context

The Class 325 is a 4 car EMU built specifically for the transportation of mail in Britain. The units have a steel bodyshell, running gear and traction package, heavily based on Class 319 EMU's. They entered service in September 1996.

3.2 The Target

The contract for these units included a requirement for achievement of 75,000 Miles Between Technical Failures Affecting Service (MBTFAS); this encompassing all Technical Failures of the units which caused a delay of five minutes or more. There were no specific remedies for failure to meet the reliability target which would therefore be treated like any other aspect of performance under the contract.

This target was set based on performance of current EMU's such as Class 319's and allowing for the removal of some sources of failure (e.g. power operated doors) and some level of engineering improvement. Reliability data for the Class 319s had been made available to ABB and they believed this target to be tough but achievable.

3.3 The Starting Point

During the design process a number of decisions were made to go for tried and tested equipment rather than more novel equipment. For example, it was decided to base the traction on Class 319 equipment rather than the more innovative Class 465 equipment. It was also decided to use an MA set rather than an alternator for similar reasons. In trying to meet the target, the project team took a number of key steps and decisions:

1 The traction package was based on the Class 319 which was performing reasonably well in service. The project team spent a significant amount of effort trying to establish the nature and cause of all the existing problems with this traction package and agreeing with GEC Traction, the manufacturer, how the existing problems would be solved.

2 The brake system and bogies were both systems which had been tried and tested on BR

3 The cab air conditioning was procured from Hagenuck, believed to be the most reliable supplier in Europe

4 The only novel equipments were an axle driven alternator, roller shutter doors, security systems and the load restraint system.

5 Major efforts were made to establish in service data for equipment and to ensure that any existing problems were eliminated. A total of 43 problems were identified and had to be investigated and solutions found.

6 Reliability risks were identified and addressed by analysis or testing. A summary of the tests is presented in Table 2.

Thus, the units incorporated as little innovation as possible, using mainly small developments of tried and tested equipment. This meant they formed a sound basis for achieving reliability.

3.4 Management of Reliability Growth

The management of reliability growth was characterised by a number of factors:

a) A FRACAS system was set up under which every failure had to be investigated and closed out. Dismissing failure as 'One Off' was not accepted and close out of failures was subject to tough peer reviews.

b) The traction system was identified as being crucial and the subcontractor (GEC) was very actively managed.

c) The customer was very good at providing information on in-service incidents and giving ABB help with investigations.

d) A large number of problems were tackled concurrently rather than only tackling a few major problems.

e) The project manager and the project team had absolute commitment to achieving the reliability target.

3.5 The Result

A graph of Failure Affecting Service against time is shown in Figure 2.

It can be seen that by the Summer of 1996 the units were achieving were achieving approximately 7,000 MBTFAS.

Analysis of a sample of 39 failures occurring at that time showed that 16 (41%) were supplier quality problems typical of the so called "infant mortality" period..

Eight (20%) of the failures were due to poor design by the supplier and of these, two should have been found by testing to show design compliance and a further three would have been found had the suppliers done overstress or environmental testing. Most of these problems were expected to be straightforward to solve.

Quality Control by ABB, in the assembly process, was responsible for only six failures; of the sort typical of "infant mortality".

ABB design was responsible for six failures all of which would be fairly easy to rectify.

Hence, although the trains were far short of the target, the overwhelming picture was that this was not because of hard to understand and fix design problems but was because of the kind of quality problems typically referred to as "infant mortality" and some design deficiencies, which might be quickly rectified.

This rapidly grew reliability, by means of some 200 modifications, to the figure of approximately 17,000 MBTFAS by February 1997. At this stage, a number of systematic failure modes were apparent, some of which were proving harder to fix than expected.

Work on identifying and solving these causes of failure continued and by August 1997 the target of 75,000 MBTFAS had been met.

3.6 Conclusion

The Class 325 reliability growth programme worked extremely well. The availability of good reliability data at the start of the contract made it possible to identify what had to be done to achieve reliability. The procurement authority was very helpful in promoting the use of low risk solutions and the ABB team had a clear plan to achieve the target.

4. CASE STUDY 3 The Class 168 Chiltern DMU

4.1 The Context

The Class 168 DMU was procured by Porterbrook Leasing and M40 Trains jointly. It was designed to be a "go anywhere" DMU with a high standard of passenger comfort and 100 MPH capability

Adtranz decided this would be the first of their new Turbostar generic concept trains and some major systems were based on generic development work carried out well before any orders were placed.

4.2 The Contract

The reliability contract required a level of reliability at least 50% better than current Class 165 trains being run by Chiltern. Adtranz were in full possession of reliability data for the current trains and, even at time of tender, had a clear plan of how to achieve that reliability.

A major driver for reliability was that the trains would be maintained by Adtranz with severe availability penalties and, essentially, fixed costs so that if components were unreliable then Adtranz would make a big loss on the maintenance contract.

4.3 The Starting Point

The selection of suppliers was based, for the most part, on a full life cycle cost assessment and suppliers were expected to validate that the product was tried and tested. Every effort was made to check data given by suppliers and references were asked for. Our door supplier (IFE)

said this was the first time they had actually known a customer take up these references and ask the previous customers how satisfied they were!

Previously unsatisfactory system concepts were avoided (e.g. split air conditioning systems) and every effort was made to ensure that intrinsically robust solutions were chosen. As far as possible we tried to buy manufacturers "standard products" although clearly they would often have to adapt these slightly for our use. Some key decisions were:-

1. Use of the MTU engine. This was the only available Euro II Specification engine. MTU guaranteed life cycle costs which were significantly lower than any competitor. Previous customers of MTU gave them an extremely good reference.

2. Direct coupling of the gearbox. This would reduce many of the problems of torsional vibration in driveshafts previously experienced.

3. Rafting of all major underframe assemblies. This would ensure that any difficult maintenance tasks could easily be done off vehicle. Not only does this greatly help availability but it also means that the maintenance should be done to a better standard.

4. Use of roof mounted electric powered Heating Ventilation and Air Conditioning (HVAC) from Air International. Adtranz had done a great deal of development work with Air International who had a very good reputation. It was decided to use electric power so as to eliminate the problems of hydrostatics in the roof and also to provide commonality with Electrostar, the Adtranz generic EMU.

5. Use of electric sliding plug doors from IFE. It is generally accepted that electric doors are more reliable than pneumatic. It was also expected that the much better fault diagnostics inherent in these doors would reduce the very high levels of door No Fault Founds.

6. Fitting of Ni-Cd batteries rather than lead acid, hence achieving better reliability and lower life cycle costs at the expense of higher first cost.

These key decisions have made us confident that the Class 168's have a sound basis from which to grow reliability.

As with the Royal Mail project every effort was made to obtain information on previous reliability problems with vehicles and ensure that they were eliminated from the Class 168 design.

Another major development in ensuring a sound basic design has been our new integrated product validation methodology. This has involved some stringent testing and verification activities.

4.4 Management of Reliability Growth

Following introduction of the fleet in May 1998, we have made every effort to grow reliability rapidly. We have had "jockeys" riding the trains so that we can get accurate reports of problems arising. All problems are logged on an Information System and every problem, even if it appears to be a one off, is thoroughly investigated. In order to close out problems the engineer must convince a sceptical panel that they have a solution. This panel includes members from Adtranz Train Maintenance Services who will have to maintain the vehicles and, hence, have a deep interest in ensuring reliability.

The train operator is committed to helping us achieve reliability and we have a very good relationship with him which is sure to help with reliability growth.

Because we manage the depot we can ensure that the reporting of work done by the depot is good and therefore will be able to clearly identify the nature of problems. This is helped by the diagnostics incorporated into many of the systems which will reduce the number of No Fault Found incidents.

4.5 The Result

At this stage it is too early to judge the effectiveness of our approach. However, the project team are working extremely hard to improve reliability. It is clear that senior personnel in both Adtranz and the customer regard reliability very seriously and that resources to solve problems are being made available. I am therefore confident that by the time of the conference in late November I will be able to report success in achieving reliability growth.

4.6 Conclusion

At this early stage it is difficult to say exactly how things are going but the indication is that all the elements are there to allow rapid reliability growth.

5 OVERALL CONCLUSIONS

Reliability growth of the Class 465s was hampered by a number of factors. The original design was heavily influenced by the customer to be very innovative and complex. The industry at this time did not have ready access to reliability data and were perhaps a bit naive in agreeing to reliability targets without having full understanding of what might be required to achieve them. The difference in perception of reliability between the Operator and the Contractor has led to some problems which are being addressed in future contracts between the two parties.

The Royal Mail project was a very successful example of reliability growth. The target was hard but clearly achievable. The procurement authority worked with Adtranz to ensure that low risk design solutions were adopted. The customer was very good at reporting defects and overall the total management of all parties was clearly directed to achieving reliability.

The Class 168 project has been good so far in ensuring a sound basis for achieving reliability by picking proven high reliability systems and suppliers and applying our new product validation methodology. The diagnostic systems incorporated in the train and the fact that Adtranz are doing the maintenance mean that data feedback is good which much improves the speed of identifying problems. These factors mean that the potential to achieve rapid growth is there and with the current high level of management commitment by all parties we are confident that growth will be rapid.

Overall the case studies support the idea that in order to achieve reliability growth a realistic target must be set and all parties must understand and be committed to it; the basic design of the vehicle must be sound and the reliability data must be adequate to allow for the rapid identification of problems.

Table 1

ADV No.	UNIT1	VEH	UNIT2	DEF DATE	CASUALTY	DELAY	CT FP	DIV LOG No	SERVICE	FAULT REPORT	ACTUAL FAULT	REMEDIAL ACTION	REP LOC	DATE STOP	RESPON SIBILITY
	465007	64765	0	01/01/96	TRUE	7 LS			07:19 GP-ON	[LOG 00012] DRIVERS BRAKE GAUGE NOT READING CORRECTLY	GP FOUND BRAKE ISOLATED	REINSTATED AND TESTED OK - LET RUN.	GP		MISC
B35436	465003	72032	0	02/01/96	TRUE	20			11.22CX-HY	[LOG 00062] DOOR TROUBLE. DRIVER LATER REPORTS ONE SET OF DOORS NOT CLOSING PROPERLY. DOORS LOOOU.	"B" DOORS UNLOCKED AND TESTED NDF	NDF	SG		NDF
	465023	72072	465019	02/01/96	TRUE	5			18.29 OT-CS	[LOG 00104] TRAIN STOPPED AT WOOLWICH ARSENAL. FOUND TWO OUTER SKINS BROKEN. MADE SAFE AND LET RUN		DIT.			UI
B35421	465177	65826	0	02/01/96	TRUE	5	C		20.13 HY-CX	[LOG 00103] FURTHER TO LOG ITEM 00099. 465177 SHOE GEAR EARTHED BETWEEN HAYS AND WWICKHAM.	LATER 22.43 HY-CX 465193 c.65842 SUFFERED THE SAME AS ABOVE.	ALL SHOE GEAR RENEWED ON BOTH UNITS.	SG		OPS
	465106	0	0	02/01/96	TRUE	10			13.08 VA-ON	[LOG 00071] DRIVER DEALING WITH DOOR PROBLEMS.		NO T+RS ATTENTION.			MISC
	465012	64820	0	03/01/96	TRUE	8			22.40 OT-CX	[LOG 00183] DELAYED AT HITHER GREEN WITH YOUTHS IN REAR CAB	CX FITTER MET ON ARRIVAL & FOUND CEILING PANEL DAMAGED,COACH LOCKED OUT OF USE.	DIT	SG		PI
	465187	65853	465050	03/01/96	TRUE	5	C			[LOG 00115] CX FITTER REPORTS SEVERAL ATTEMPTS MADE TO COUPLE UNITS DUE TO DIFFERENCES IN COUPLING HIGHTS. BOTH UNITS REQUESTED TO DEPOT FOR COUPLING EXAM.		DIT.			UI
	465189	72977	0	03/01/96	TRUE	12			10.23 ON-VA	[LOG 00149] DELAYED AT KENT HOUSE WITH DOOR TROUBLE.	VA FITTER STATES "B" DOORS NOT CLOSING FULLY.	UNIT REBOOTED AND TESTED DOORS OK. L/R/F/F/R.	VA		E
	465002	0	0	04/01/96	TRUE	18 LS			16+13 SG-CX	[LOG 00259] ETC OPEN AND 1ST AID BOX MISSING.	CAME TO SLADE GREEN	RENEWED	SG		PI
	465012	70251	0	04/01/96	TRUE	5			18.28 SG-CS	[LOG 00271] DELAYED AT WOOLWICH	DRIVER MAKING SAFE A BROKEN LARGE SIDE LIGHT	DIT			UI
	465020	64828	0	04/01/96	TRUE	C			19.25 OT-CX	[LOG 00280] DEFECTIVE W/WIPER	WENT GROVE PARK FOUND BLADE DEFECTIVE	RENEWED.	GP		E
B35634	465042	64800	0	04/01/96	TRUE	12 LS			07-30 SG-CYD	[LOG 00201] REPORTED AS CAB DOOR SLOW TO CLOSE UNIT DETACHED	CAME TO SLADE GREEN AND DOORS TESTED	NDF	SG	04/01/96	NDF
	465158	0	465177	04/01/96	TRUE	6			08.44 CX-GV	[LOG 00230] REPORTED FOR LOSS OF TRACTION CURRENT FROM DARTFORD TO NORTHFLEET.UNITS REQUIRE SHOE GAUGING.	465156 CAME TO SLADE GREEN AND CLEARED. (465177 INVOLVED IN LOSS OF SHOE AT W/WICKHAM 2/1/96 THEREFORE SHOULD NOT NEED GAUGING ?)	465177 DIT			UI

SYSTEM	TEST
Roller Shutter Door	Static Load, Climatic, Endurance, Overstress
Axle Driven Alternator	Climatic, Endurance, Type, Overstress, Vibration
Payload Floor	Endurance, Overstress
Windscreen Wipers	Endurance, Overstress
Shoegear	Climatic, Endurance
Air Conditioning	Vibration, Endurance, Overstress
PIR	Functional
Smoke	Functional
Brake Module	Overstress
Panto Module	Overstress
Compressor Motor	Type, Overstress
WSP Probe	Vibration
Drivers Seat	Overstress
Traction Motor	Overstress
MA Set	Overstress
Keysafe	Endurance
Restraints	Functional
Sunblind	Endurance
AC/DC	Endurance
EMU/LOCO	Functional
Water Ingress	Type
Horns	Type

Table 2 - Reliability and Overstress Testing Programme

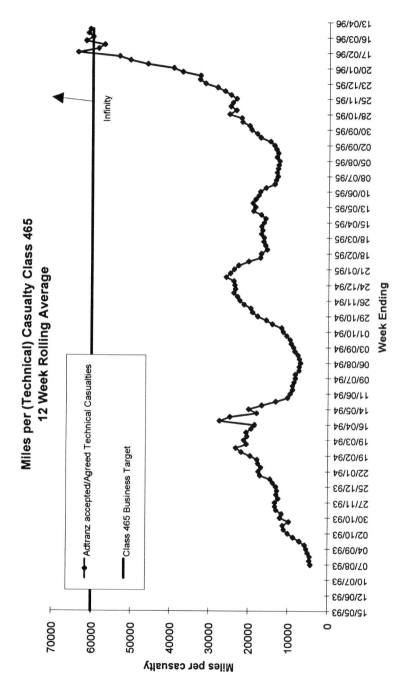

Figure 1

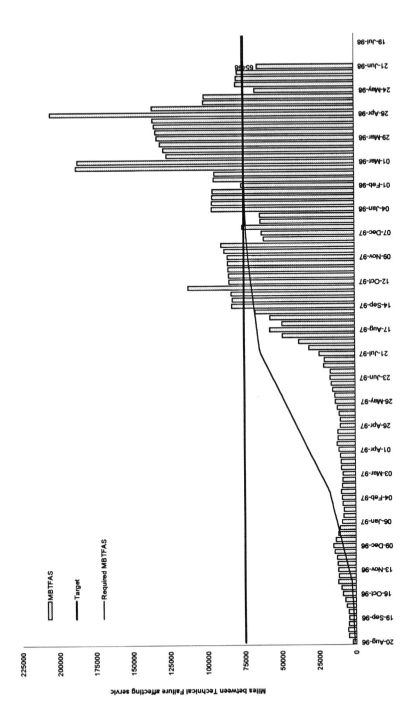

Figure 2 Reliability Growth For Class 325

C552/016/98

The safety consideration of introducing new rolling stock on an adjacent railway system

W HOSKINS BSc, MSaRS, AMIEE, AIL
AEA Technology TCI Rail, London, UK

1. SYNOPSIS

As part of the Jubilee Line Extension Project (JLEP), new rolling stock has been introduced onto both the existing and extended parts of Jubilee Line. The operation of these trains needed to be carefully considered with regard to the possible effects on the neighbours - particularly Railtrack zones which operate adjacent infrastructure on sections of the North London Line and the Chiltern Line.

Transportation Consultants International (TCI) was called in to help the Jubilee Line Extension Project demonstrate to Railtrack that there would be no negative impact caused by the new trains. A multi-stage process was then started with the goal of obtaining Railtrack's agreement to operate the trains.

The work involved modelling the electromagnetic coupling between the railway systems, performing tests to validate the modelling and separately measuring emissions from the new rolling stock itself. The work had to be performed with minimum disturbance to Railtrack operations.

This paper summarises the issues, strategies and work performed to obtain Railtrack's agreement, but avoids overly detailed technical discussion.

2. BACKGROUND

The Jubilee Line Extension extends the London Underground Ltd., Jubilee Line from Green Park through South London and Docklands and east to Stratford. In 1994 an adjacent section of Railtrack infrastructure, the North London Line between Channelsea Junction and Silvertown, was realigned to provide space for the Jubilee Line. This line was also resignalled by JLEP on behalf of Railtrack.

The existing above ground section of the new LUL Jubilee Line already runs parallel to the Railtrack Chiltern Lines between Finchley Road and Wembley Park. Here, the Down

Line of the Metropolitan Lines is situated between the Jubilee Line tracks and the Up Main of the Chiltern Line, with the Down Main of the Chiltern Lines being beyond the Up Main.

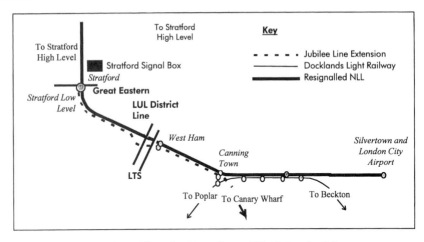

Figure 1 The Jubilee Line Extension and North London Lines

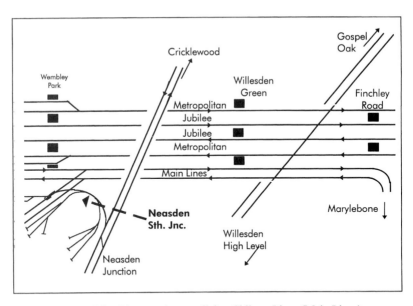

Figure 2 Jubilee Line running parallel to Chiltern Lines (Main Lines)

This paper explains the approach used by TCI in assessing the Electromagnetic Compatibility of the existing North London Line and Chiltern Line signalling in the presence of the new JLE Tube Stock, signalling, communications, and all infrastructure and stations associated with the Jubilee Line Extension from Stratford to Canning Town.

3. NORTH LONDON LINE AREA: THE APPROACH

3.1 Approach

The Jubilee Line Extension project was clearly committed to ensuring the safe operation of their new services. This was echoed by Railtrack which, whilst accepting London Underground's commitment, could not be content with leaving such safety issues entirely to another party, and needed to be involved in the process to satisfy their own safety culture.

The initial approach was to develop a strategy, acceptable to Railtrack, whereby trains could be introduced into operation and tested without interrupting Railtrack's operations. This approach consisted of determining the effect on the parallel NLL infrastructure of running a single (initially) 96TS test train along the JLE track, which was located above ground between Stratford Low Level and Canning Town. The study would then be extended to cover the test operation of the fleet of 96TS trainsets on the JLE.

Essentially, the approach taken for the North London Line consisted of:

- Determining any hazards which existed
- Showing that it was safe to operate a single train on the JLE railway
- Measuring train emissions and gradually introducing, in phases agreed with Railtrack, further 96TS trains.

A Hazard and Operability Study (HazOpS) was conducted in order to identify and categorise the potential hazards and, to ensure that all elements of the JLEP and the NLL, in the area where they run parallel, were identified and recorded. In support of this, a general route survey was performed to note any recent changes to the railway infrastructure and provide supplementary information to signalling and electrification plans.

The equipment considered and concluded to be subject to no significant risks included: *Main Colourlight Signal, Position Light Shunt Signal, Junction Indicator, Stencil Route Indicator, CCTV (Security), AWS, Train Describer, Line Circuits, Tail cables, Signalling Power Supplies, Traction Control Circuits, SSI, Point Clamplocks, Staff Warning Systems, Patrolmen's Lockout Systems.* Reed track circuits were also considered but, due to their location were well beyond the zone of influence.

Items therefore considered to be most at risk included Telecommunications systems, and TI21 and HVI track circuits. The Jubilee Line's own Telecommunications equipment was subsequently agreed to be operating in licensed frequency bands which would not affect Railtrack equipment.

The Jubilee Line Extension is not directly electrically connected to either the North London Line or the Chiltern Lines. However, it was determined that electrical interference could still be coupled by the mechanism of electrical induction and a coupling model was developed by TCI to determine the potential levels and effects of EMI transferred in this manner.

3.2 The problems associated with modelling

Modelling is a substitute, at least in part, for *measuring*. In the case of this study it was the impracticality of performing sufficient measurements without severely disrupting Railtrack operations which dictated the need to use an accurate coupling model.

However, the onus was still on the Jubilee Line Extension Project to in some way demonstrate, with supporting measurements, that from Railtrack's point of view the 96TS trains were safe.

There was therefore not only the need to develop the model but also to ensure that it could be shown to be accurate and acceptable to Railtrack.

4. THE COUPLING MODEL

The assessment of potential effects on HVI and TI21 track circuits was based on establishing the coupling mechanism between the JLE and track circuits on Railtrack's infrastructure. The Carson-Pollaczek method, used in previous studies [e.g. reference 1], was applied in the calculation. The objective of this work was to refine the representation of the impedance between two sets of running rails.

For example, any interference currents drawn by a train on the JLE system may induce voltages in the adjacent NLL TI21 track circuits. This relationship can be defined as a coupling factor in terms of mV/m/A. This coupling factor depends primarily on the topology of the circuit and the current distribution both in the rails and flowing in the earth.

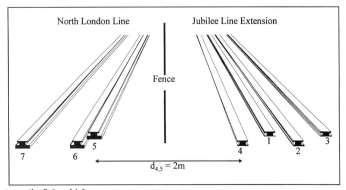

Running rails: 7,6 and 1,3

Figure 3 Topology of NLL-JLE conductors

To validate the model it was proposed that a test should be performed on JLE using the two roads between West Ham and Canning Town. This would require current to be injected down one track and the level of induced current measured in the other. Such a test would show that the model produced acceptable results whilst, by virtue of the fact that it did not involve making electrical connections to Railtrack lines, would minimise disruptions. The method was described to Railtrack representatives who suggested that before such validation tests could be permitted, a simple check should be performed to ensure that test currents injected onto the JLE would not themselves interfere with Railtrack's infrastructure. This check became known as the calibration test.

4.1 Test results

The calibration test involved injecting a high current signal (at 12A, the signal was beyond any level the 96TS train would be likely produce at TI21 frequencies) into a JLE conductor rail and measuring the voltage coupled onto a TI21 track circuit receiver. The levels recorded on the exposed section were within the amount deemed acceptable as levels of background noise [Reference 2].

This established safe limits for the required validation tests but did not in itself demonstrate safety as only a limited section of the railway was exposed to interference currents. Having established these safe limits, tests were then performed on the Jubilee Lines to obtain the data necessary to validate the impedance model. Again, a 12A source was used to inject current (the calibration test having already shown that this would not disturb the exposed section of the NLL infrastructure).

5. TI21 TRACK CIRCUITS (NLL)

The validated case showed that modelling could be reasonably applied to determine currents induced from one track to another. The model was then applied to the specific arrangement that exists between the Jubilee Line and North London Line systems. There is fundamentally no difference in the coupling mechanisms which act between any of the 4 tracks in this area and it was argued that what is true between the Jubilee Line tracks remains so between Jubilee and adjacent NLL tracks. Allowances were made in the model for the different topology and the inclusion of the metallic fence which separates the two infrastructures for much of the area of interest.

Parameters, such as the proportion of the earth current, depend on many factors including the earth resistivity, system configuration and weather conditions. Similarly the distribution of the return current in the returning rails depends on the system topology. It was found from the calculations that the transverse voltage level is very sensitive to the variation in the earth current proportion and also to the current distribution in the rails. Inconsistent measurement were obtained on different days confirming the hypothesis that earth resistance is the most sensitive parameter.

The purpose of the model was, however to show not an exact value (which in any case would be varying with weather on a daily basis) but a likely range which included using reasonable worst case conditions.

Using the validated model with reasonable worst case parameters, coupling factors were produced in terms of the level of interference current in mA/A (that is, **mA** in the NLL TI21 track circuit per **1A** of the 96TS current for the full length of exposure). Two states were considered in order to calculate values for different conditions. These were a fault on the conductor rail of the JLE (positive or negative rail) and the effect of a metallic fence which separated the two railways. The results for the different combinations of these two states are shown in Table 1.

Infrastrucure considered	Fault	Fence	Fault	Fence	Fault	Fence	Fault	Fence
	X	X	X	✓	✓	X	✓	✓
Coupling (mA per A)	9.8		8.3		16		14	

Table 1 NLL TI21 current in mA per 1A interference on JLE

5.1 Confirmation of model against coupling measured between JLE and NLL

During the calibration test on the JLE tracks, the background noise was measured as a transverse voltage across the positive conductor rail and the running rails. It was observed that this background noise contained a level of TI21 frequencies, induced in the JLE tracks from the adjacent TI21 track circuits. One example is shown in Figure 4 for a TI21 frequency of 2610 Hz.

The calculations of the inductive coupling are applied to the system in the same fashion but in this case the inducing circuit is the NLL TI21 track circuit while the affected circuit is the JLE positive conductor in series with the running rails. In order to calculate this level the following information is required:

- The operating current of the TI21 track circuit
- The TI21 track circuit length (effectively the exposure length)

For a distance $d_{4,5}$ =2 metres and a current of 1 Amp flowing in conductor number 7 and returning in conductor number 6 at a frequency of 2610 Hz [1] the calculated transverse voltage $V_{3,4}$ is 566 mV / A / km.

The exact parameters which should be used for the modelling could not be determined as it was not possible to distinguish which TI21 track circuit had produced the signal. The following parameters were therefore taken for this check:

- TI21 Current: 0.3 A (maximum operating value)
- Length of track circuit: 0.3 km (typical for TI21 track circuits of this frequency).

---

[1] The nominal centre operating frequency (f_d) of the TI21 track circuit is 2593Hz. The peak amplitude of the sideband was taken at 2610Hz from the signal trace.

The calculated voltage $V_{3,4}$ developed is therefore given by: $V_{3,4}$ = 566 x 0.3 x 0.3 = **50.9mV**

As seen in Figure 4 the level of the transverse voltage measured was approximately 50mV.

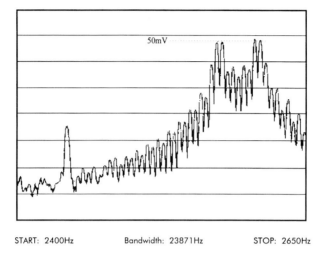

START: 2400Hz   Bandwidth: 23871Hz   STOP: 2650Hz

Figure 4 Trace of TI21 signal detected on JLE

The result is indicative of a high level of accuracy showing the model is suitable for its intended purpose.

6. HVI TRACK CIRCUITS (NLL)

6.1   HVI Track Circuit Operation

A driving pulse voltage applied to an HVI receiver will result in a current flowing in each of the two relay coils (Figure 5). The closure of the relay front contacts will depend on both these coil currents (I_1 and I_2) passing threshold values and entering what is referred to as the pick up 'V' curve (see Figure 6). This depends on the energy in the driving pulse voltage[2]. The front contacts will open when one or both the coil currents falls outside the drop away 'V' curve.

---

[2] If the *mmfs* in limbs 1 and 2 are roughly balanced (corresponding to I_1 and I_2 being inside the pick up 'V' curve), the flux in the third limb will be sufficient to pick up the armature and consequently close the front contacts of the track relay. If the *mmfs* in the two limbs are unequal (corresponding to I_1 and I_2 outside the pick up 'V' curve), a circulating flux will be developed in limbs 1 and 2 at the expense of the third limb. If the reduced flux in the third limb is below the threshold level the armature and front contacts will not operate.

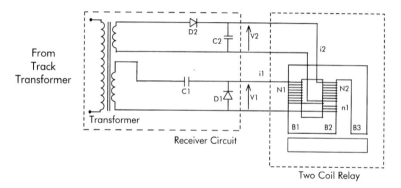

Figure 5 Principle of operation HVI Receiver circuit

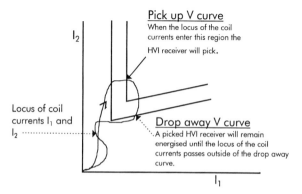

Figure 6 HVI pick up and drop away 'V' curves

It must be borne in mind that a relay pick up criterion based on $\delta i/\delta t$ alone is inadequate, since it is the energy in the resultant voltage pulse[3] that determines whether the relay will pick up or not. Therefore, the duration of the $\delta i/\delta t$ event must be specified. A $\delta i/\delta t$ with meaningful duration is considered to be equivalent to a pulse voltage with an amplitude and pulse width[4].

---

[3] Given by M* $\delta i/\delta t$, where M denotes the mutual coupling between all of the conductors present. In this case, there are 8 conductors (see Figure 3) and the mutual coupling is therefore calculated by an 8 x 8 matrix operation.

[4] Laboratory tests carried out at GEC-Alsthom on a HVI receiver found that pulse voltages with narrow pulse widths required a much higher amplitude to pick the relay than those pulse voltages with a larger pulse width. Of course, this argument is only true up to the saturation point of the track transformer. On saturation the secondary voltage collapses.

6.2 High Voltage Impulse (HVI) Track Circuit Modelling

Impedance between the Jubilee Line and the North London Line is considered in the same fashion as for the TI21 track circuit model. The mutual coupling model of the JLE and NLL tracks includes the following items:

- An 8 x 8 mutual coupling model (as validated for the TI21 case) linking the 8 conductors (4 LUL rails, 3 NLL rails and a fence separating them)
- The self impedance of the 4 LUL rails.
- The self impedance of the 3 NLL rails.
- A stationary train standing at the transmitter end of a 200 metre long HVI track circuit.
- The HVI receiver.

The HVI receiver model includes:

- The track transformer - This is basically a pulse transformer, and has been modelled using a saturable non-linear B-H characteristic obtained through laboratory measurement at the track circuit manufacturer's laboratory at GEC-Alsthom, Trafford Park, Manchester.
- The receiver circuit - This consists of a three-winding transformer and two capacitor charging circuits (one for each of the two relay coils).
- The track relay - This is a complex three limbs and two coils relay. The model takes account of the mutual coupling between the two coils.

6.3 Simulation of high energy transient

The HVI and impedance models permit a calculation to be made of the transient current which would actually be required in order to cause the HVI track circuits on the NLL to become energised. This is not meant to represent an achievable current, but indicates how much current is required because of the high impedance between the two rail systems.

To determine this figure an unrealistically high level of dc current at X Amps is simulated between the +420V and -210V rail. The current is then interrupted and allowed to fall to zero in 10ms (typical fall time). This gives a rate of change of JLE train current at $X \div 10$ Amps / ms. A sudden and direct transient of this nature is that which is most likely to affect an HVI track circuit as it has no noise associated with it, noise ordinarily desensitising the receiver. A true transient from a 96TS train, apart from being very much smaller, would naturally contain noise elements.

Upon a current interruption a longitudinal voltage is induced on to each conductor through the mutual coupling mechanism. The amount of voltage is dependent on the impedance between the two systems which is determined from the impedance model. The HVI receiver input voltage is the induced transverse voltage; the difference between the two longitudinal voltages on the NLL HVI track circuit conductors.

This voltage is then used to drive the receiver model. This was carried out in a dynamic transient simulation where all these events happen simultaneously.

The combined *theoretical* JLE traction current is gradually increased until the HVI track relay's front contacts are just closed. This point is indicated on the 'V' curve plot where the locus of I_2 against I_1 is just inside the pick up 'V' curve (see Figure 6).

This exercise was carried out with and without the fence having a proper return to ground.

6.4 Theoretical maximum permissible currents

The determination of the values at which the HVI track relay contacts close defines the theoretical maximum permissible limits to emissions on the Jubilee Line. The current would need to change from its nominal level of X amps to zero in a typical fall time of 10ms. It was determined that the values of the starting current X that would actually be required (shown in Table 2 below) can not be achieved on the JLE infrastructure and therefore the possibility of affecting the HVI track circuits was negligible.

Scenarios considered		Sustained $\delta i/\delta t$ required to cause Wrong Side Failure on adjacent NLL infrastructure	Traction current on JLE required to energise HVI track circuit on NLL
Fence	Train (NLL)		(Typical fall time of 10ms) X Amps
X	✓	10,000 A/ms	100 kA - *Not possible*
X	X	24,800 A/ms	248 kA - *Not possible*
✓	✓	30,000 A/ms	300 kA - *Not possible*
✓	X	60,000 A/ms	600 kA - *Not possible*

Table 2 Theoretical Levels required to affect HVI track circuits

7. TRAIN TESTS

Having ascertained the safety of operating a single train on the JLEP test track area, the question still remained as to how full fleet operation may impact on EMC. One of the most important features, certainly in terms of EMC impact, of the 96TS trains being introduced by LUL was the fact that regenerative braking was to be employed. Railtrack, with no previous history of actual usage, were understandably concerned and this was a feature specifically investigated. Figure 7 below is just one example of measurements taken during operation on the JLEP test track (Note substation harmonic frequencies at 1200Hz and 2400Hz).

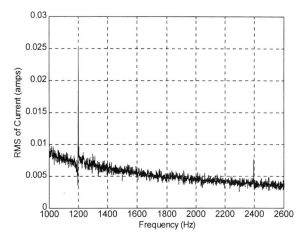

Figure 7 Frequency Assessment of regenerative braking on JLE
'Sink' train accelerating from 15kph, 'Source' train braking from 60kph.

The maximum time averaged emission of a single train seen at TI21 frequencies, i.e. between 1682Hz and 2610Hz, ignoring substation related emissions at 2400Hz, was approximately 6mA. This was the value used for analysis of the effects of coupled emissions on TI21 track circuits.

The results confirmed that Railtrack infrastructure could not be considered at risk even during full fleet operation on the Jubilee Line.

8. THE CHILTERN LINE AREA

The Down Line of the LUL Metropolitan Lines is situated between the Jubilee Line tracks and the Up Main of the Chiltern Line, with the Down Main of the Chiltern Lines being beyond the Up Main. This essentially means that, with the tracks further apart than in the North London Lines case, the amount of inductive coupling will be less significant.

However, detailed assessments were performed to ensure that 96TS fleet operations would not interfere with Railtrack operations. This comprised of:

- A Hazard and Operability Study which identified the Railtrack signalling and telecommunications equipment which may be at risk.
- Site inspections which verified equipment and confirmed the lack of deliberate conductive paths between the LUL and Railtrack tracks in the area under consideration
- Examination of drawings and technical records of the relevant signalling and telecommunications infrastructure (LUL and Railtrack).

- A comparison of the Chiltern Lines with the GWML and North Pole Depot area where Reed and TI21 track circuits are in use adjacent to separate tracks used by rolling stock producing EMI at Reed frequencies.
- A comparison with the Stratford-Canning Town area where the North London Lines and Jubilee Lines run parallel and the 96TS trains were, by this stage, in daily use.

The main concern centred around a number of Reed Track Circuits located at Neasden South Junction. Reed track circuits are considered more susceptible to interference than the TI21 or HVI types. Replacing these track circuits would have been a major exercise with significant design and implementation costs.

However, investigation by TCI traction and signalling experts were able to show that these track circuits were not considered at risk due to LUL Jubilee Line trains and an extremely expensive replacement exercise was thereby avoided.

9. CONCLUSIONS

- The evidence of the two studies presented to Railtrack resulted in the issue of letters of no objection to operation of a fleet of 96TS trains on sections of the London Underground Jubilee Line. The two organisations, both intent on ensuring safe train operation, were able to agree an approach and subsequently the details and conclusions of specific safety studies and tests.
- Modelling has been shown to be a practical approach for investigating EMC safety issues, but still needs appropriate validation with cost and program implications.
- The studies showed that expensive track circuit replacements around the Neasden South Junction area could be avoided. This demonstrated that considerable financial savings can be made when the nature and level of risk associated with particular hazards is well understood.

10. REFERENCES

1. Mellitt B, Allan J, Shao Z Y, Johnston W B, Goodman C J, "Computer Based Methods for Induced Voltage Calculations in AC Railways", IEE Proceedings, January 1990.
2. Train Detection Handbook, GK/RC0761, June 1996.

11. ACKNOWLEDGEMENTS

Mr Mike Evans, EMC co-ordinator for JLEP.

Dr Rafat Kadhim and Dr Steve Goh, TCI.

© With Author 1998

The Institution of Mechanical Engineers is a leading forum for the exchange of knowledge and expertise in the field of mechanical engineering.

A wide range of events is organized by the IMechE, to which all are welcome to attend. For further information about IMechE Conferences, Seminars, Workshops, and other events please contact us for further information.

Visit our website	www.imeche.org.uk
Telephone	+44 (0) 171 222 7899
Fax	+44 (0) 171 222 4557
Or write to	Conferences and Events Institution of Mechanical Engineers 1, Birdcage Walk London SW1H 9JJ

We look forward to seeing you at our future events.